Project Evaluation: Essays and Case Studies

Volume II

Carl D. Martland

**Union Station, Toronto:
Modern Technology inside a Renovated Historic Structure**

Project Evaluation:
Essays and Case Studies

Volume II

Comparing Economic and Financial Impacts
Over the Life of Proposed Infrastructure Projects

© Carl D. Martland
Sugar Hill, New Hampshire
October 2016

This book is based primarily upon materials prepared between 1997 and 2010 by Carl D. Martland for *1.011 Project Evaluation*, a required course within MIT's Department of Civil & Environmental Engineering that he designed, developed, and taught for many years. It is structured to be of interest to anyone interested in infrastructure systems, especially engineers, planners and managers who design, build and operate such systems. The book may also be of interest to students in planning or engineering who are interested in transportation, water resources, energy, city planning, or real estate development.

In 2012, John Wiley & Sons, Inc. published a 500-page textbook *Toward More Sustainable Infrastructure: Project Evaluation for Planners and Engineers* that was also authored by Mr. Martland and based upon the same course materials. That book, which was designed and formatted as a standard textbook for an undergraduate class, includes many more examples, hundreds of problems for students, an additional chapter on project management, and several open-ended case studies that can be used for class assignments. Instructors who purchase that textbook and assign it for their own class can obtain the textbook, a teaching guide, solutions to problems, and related power point presentations from Wiley.

Additional materials related to *1.011 Project Evaluation* can be obtained on-line from MIT's Open Courseware website, which can be accessed from MIT's homepage (www.mit.edu).

Cover design, layout, and editing by Carl D. Martland
All photographs by Carl D. Martland unless otherwise noted
Published on-line with CreateSpace.Com
Available directly from CreateSpace.Com or Amazon Books

Cover Photo: Rebuilding a section of I90 west of Buffalo.

The Claiborne Pell Toll Bridge
Crossing Narragansett Bay between Jamestown and Newport, RI

Project Evaluation: Essays and Case Studies

Contents: Volume I

Preface vii

Essays in Volume I

Introduction
- Toward More Sustainable Infrastructure: Better Projects and Better Programs — 1
- Infrastructure Projects and Programs — 2
- Evaluating Infrastructure Projects — 5
- Infrastructure, Cities, and Civilization — 7
- Where Do Projects Come From? — 8
- A Framework for Project Evaluation — 8
- Essays and Case Studies — 12

Basic Economic Concepts
- Introduction — 14
- Supply, Demand, Equilibrium — 15
- Pricing — 20
- Productivity — 23
- Measuring and Improving the Economy — 25
- Trade — 27
- Making Decisions: Utility and Sunk Costs — 30
- Summary — 31

Public Perspectives: Economic, Environmental and Social Concerns
- Overview — 34
- Benefit/Cost Analysis — 35
- Economic Impacts: Measures Related to the Regional or National Economy — 37
- Environmental Impacts — 37
- Social Impacts — 43
- Safety and Security — 46
- Summary and Discussion — 47

Comparing Strategies for Improving System Performance
- Introduction — 48
- Discounting and Net Present Values — 48
- Measuring Cost Effectiveness — 49
- Using Weighting Schemes in Multi-Criteria Decision-Making — 49
- Seeking Public Input — 50
- Summary — 51

Public Private Partnerships
    Introduction   53
    Reasons for Considering a Public Private Partnership   54
    Principles of Public Private Partnerships   57
    Creating a Framework for a Partnership   58
    Determining How Much to Invest in a PPP   60
    Example: Using a PPP to Maximize a City's Ability to Undertake Projects - Toronto's Highway 407   61
    Example: Using Public Investment to Stimulate the Economy - Investment in off-shore oil exploration by the Province of Newfoundland and Labrador   62
    Summary   63

Evolution of Infrastructure-Based Systems
    Introduction   65
    Stage I Technological Experimentation and Demonstration   67
    Stage II Widespread, Uncoordinated Implementation   68
    Stage III Development of Systems   68
    Stage IV Consolidation and Rationalization   69
    Stage V Technological and Institutional Advancement   69
        Example: New Technology Can Revive Old Methods – Wastewater Treatment in San Diego, California   70
        Example: Replacing Old Infrastructure with More Effective Systems – Wastewater Treatment in Kaukauna, Wisconsin   70
    Stage VI Responding to Competition   71
        Examples: Competition Faced by the Airline Industry   72
    Stage VII Mitigating Social and Environmental Impacts   72
        Example: Evolution of London and the Thames Embankment   75
        Example: Building New, More Sustainable Infrastructure – One Bryant Park   76
    Stage VIII Retrenching and Obsolescence   77
    21st Century Challenges: Sustainable Infrastructure   77
    Summary   79

Case Studies in Volume I

The Franconia Notch Parkway   81

The Panama Canal
    Early Routes Across the Isthmus   86
    The Panama Railroad   87
    The French Effort   89
    The U.S. Effort   90
    Transfer of the Canal to Panama   94
    Issues for the 21st Century   94

Pearl River Delta: 'More than a Bridge"
    Background   98
    How the Team Did Its Work   100
    Lessons Learned from the Pearl River Delta Study   101

Scenario Planning at Southern California Edison	103
Financing a Bridge Project	105
Overview of Options for Financing a Bridge Project	105
Can the Bridge Be Justified as a State Project?	107
Could a Private Bridge be Financed with Tolls?	108
Should the Bridge be Built as a Public Private Partnership?	108
Public Incentives for Low-Income Housing	110
The Sheffield Flyover, Kansas City, Missouri	112
Overview	112
Results	114
Lessons Learned	114
Skyscrapers and Building Booms	116
Evolution of the U.S. Rail System	119
Overview	119
Introduction of Railroads in the Early 19th Century	119
Problems Emerge: Accidents, Greed, and Corruption	120
Monopolistic Excess and Regulation of Railroads	121
Deregulation of the Railroads	123
Technological Innovation	123
Summary	125
Lessons from the History of Railroads	125
Rehabilitating Newark's 19th Century Brick Sewers	126

Project Evaluation: A Few Final Thoughts   128

Further Reading   130

Project Evaluation: Essays & Case Studies

Contents: Volume II

Preface vii

Essays in Volume II

Introduction 1

System Performance
Performance of Infrastructure-Based Systems 3
System Cost 4
Profitability, Breakeven Volume, and Return on Investment 12
Service 15
Capacity 16
Safety and Security 16
Cost Effectiveness 17
Summary 18

Equivalence of Cash Flows
Introduction 20
Time Value of Money 22
Equivalence Relationships 23
Continuous Compounding: Nominal vs. Effective Interest Rates 29
Financing Mechanisms 31
Summary 38

Choosing a Discount Rate
Introduction 39
Profits and Rate of Return vs. Net Present Value 39
Leveraging and Risk 40
Factors Affecting the Discount Rate 42
Choosing a Discount Rate: Examples 48
Dividing Up the Cash Flows of a Major Project 49
Summary 50

Financial Assessment
Introduction 52
Maximizing Net Present Value 52
Importance of Project Life 53
Does Discounting Ignore Future Catastrophes? 54
Return on Investment and Internal Rate of Return 55
External Rate of Return 56
Constant Dollar vs. Current Dollar Analysis 58
Choosing Among Independent Investment Options 58
Choosing Among Mutually Exclusive Projects 60
Dealing with Unequal Lives of Competing Projects 61
Splitting a Project into Pieces for Different Parties 63
Summary 64

Rules of the Game: Taxes Depreciation and Regulation
    Introduction   66
    Depreciation and Taxes   67
    Income Taxes   71
    Land Use Regulations   75
    Building Codes and Other Safety Standards   77
    Environmental Regulations and Restrictions   77
    Summary   78

Dealing with Risks and Uncertainties
    Introduction   81
    Example: Dealing with Risks and Uncertainty in a Toll Road Project   82
    Using Analysis to Understand Risks   83
    Probabilistic Risk Assessment   84
    Performance-Based Technology Scanning   93
    Summary   95

Case Studies in Volume II

An Engineering-Based Service Function for Bus Operations   97

Capacity of a Highway Intersection   99

Canal Projects in the Early 19th Century
    Introduction   102
    Service and Capacity of a Canal   104
    Canal Competitiveness   107
    Engineering-Based Cost Model for a Canal   107
    Evaluating the Canal Project   109
    Examples of Canals   110
    Epilogue: Canals vs. Railroads   111
    Lessons for Other Infrastructure Projects   111

Building an Office Tower in Manhattan   113

Multiple Internal Rates of Return for a Stadium Project   116

Public Incentives for Low-Income Housing   120

Financing a Bridge Project
    Overview of Options for Financing a Bridge Project   121
    Can the Bridge be Justified as a State Project?   124
    Could a Private Bridge be Financed with Tolls?   125
    Should the Bridge be Built as a Public Private Partnership?   125

Using a Probabilistic Model to Investigate Financial Risks   127

Applying Performance-Based Technological Scanning to Intercity Passenger Transportation
    Competition for Intercity Passenger Services   130
    The Utility of Time   131

A Preliminary Model of Passenger Utility	132
Estimating Mode Shares	135
Implications for Carriers and Terminal Operators	137
Implications for Project Selection and Projects Evaluation	137

Reducing Risks Associated with Grade Crossing Accidents   131

Appendix

Equivalence Factors for Selected Discount Rates   141

**Wetlands protected by the Nature Conservancy
Cape May, New Jersey**

Project Evaluation: Essays & Case Studies
Volume II

Preface

Motivation

This book contains essays and case studies that are based upon materials that I prepared for "Project Evaluation", which I designed and taught for more than ten years as one of the required subjects in MIT's Department of Civil & Environmental Engineering. The subject was designed to fill a void in the education of civil engineering students, namely an understanding of why major infrastructure projects are undertaken, how they are structured and evaluated, and how they are financed. These topics, which naturally are of central importance to civil and environmental engineering, are related to, but certainly not central to micro-economics, the subject that was previously required for civil engineering undergrads at MIT.

Micro-economics is an interesting and challenging field, but it tends to ignore or brush quickly over some of the central issues in designing and developing infrastructure projects. Where should a project be located? When should it be built? Can it be developed in phases, so that capacity can be added only when and where it is needed? For engineers, planners, and entrepreneurs, these are critical questions. Those who want to be engineers, planners or entrepreneurs must learn how to balance current vs. future costs and benefits, and they must be able to understand and respond to the many factors that influence the pace and location of development. In particular, they must understand the time value of money, the equivalence of cash flows, and the effects of risk and inflation on discount rates and the attractiveness of projects. These are all central topics in engineering economics, but they are largely or entirely absent from the standard introduction to micro-economics. As I tell my students, economics is a bit too close to the Twilight Zone – "a dimension neither of space nor time".

A second concern with micro-economics is that many of the most interesting concepts are extremely difficult to apply without making assumptions that, to an engineer or planner or entrepreneur, seem to be simplistic or heroic or merely untenable. An engineer is likely to treat with suspicion any proposition that begins with "given a cost function" or "given a production function" or "given supply and demand curves". Where do these functions come from? How are these curves calibrated? Some economists have gone to factories and rail yards and studied the inputs and outputs actually required for the various possible means of production. More commonly, economists have relied on statistical techniques to calibrate functions that certainly appear to be very complex to the student (or to the reader of a journal article), but that in fact are a quite simple portrayal of costs or production or demand based upon analysis of what has happened in the past. For many purposes, notably many kinds of policy analysis, econometric modeling and economic theory provide useful insights, but when considering major projects, engineers, planners, and entrepreneurs are more concerned with what can be done in the future than with what was done in the past. New technologies, new designs, changes in relative costs of inputs, and many other factors will influence what will be possible or desirable to do in the future. Someone, presumably the engineers and the planners, will have to figure out what can be done and convince others that it should be done, tasks that require creativity and judgment as well as an understanding of complex systems and methodologies.

While I understand the argument that undergraduates should learn the basics of their field and that they should discipline their mind through thorough rigorous examination and understanding of a complex, intellectually stimulating subject, my personal experience suggests that students require the stimulation of real situations to truly understand the concepts that we try to teach them. Moreover, it is possible to over-emphasize methodologies and theories while doing little to encourage independent thought and initiative. Thus, in designing my class on project evaluation, I included case studies, open-ended problem sets, and a term project in which the students investigated projects of their own choosing. I had students complete some exercises from an engineering economics textbook, but I was much more interested in how well they could apply the methodologies and ideas in analyses and interpretations of realistic problems.

At this point I should add a short note on my background. As an undergrad I studied math, but lost interest as the theory deepened and the potential applications receded. As a senior and then in graduate school, I shifted to studying what was just beginning to be called "urban systems", but eventually ended up writing a thesis on rail freight system reliability. For the next 35 years, I remained on the research staff at MIT, supervising many research projects that were funded by the rail freight industry – an experience that forced me and my students to pay great attention to detail and to reality. In effect, we spent several decades working with rail researchers and field personnel to understand and improve the cost functions and production functions related to various categories of rail freight. Over this period, the rail industry transformed itself from a nearly bankrupt, over-built and under-maintained system into a thriving, streamlined system with more trains, longer and heavier trains, heavier loads, and more efficient equipment and facilities. The industry had little to spend on research, so it went to great efforts to focus that research on areas where there would be a payoff. Participating in this research proved to be an outstanding way to understand the functioning of an extremely complex, long-lived system as it was updating its infrastructure and equipment to serve new markets.

During my research career, I described much of what I did as being some sort of engineering economics. Several aspects of engineering economics were absolutely critical:

- Net present value and equivalence of cash flows: the ability to compare cash flows over long time horizons for multiple alternatives, often in an attempt to understand the potential for new technologies or operating strategies.
- Engineering-based cost and performance functions: the ability to structure detailed cost and performance functions that captured the relevant aspects of the technologies and operations that were of interest.
- Probabilistic analysis: the ability to include probabilistic features when structuring cost and service functions.
- Identification of key factors: the use of financial analysis, scenarios, and sensitivity analysis to identify the most important factors affecting a project, the use of new technology, or the choice of operating or marketing strategies.
- Approximation: appreciation of the fact that it is seldom necessary to obtain precise results in order to reach solid conclusions.
- Structuring and interpreting results: recognizing that lack of consensus regarding objectives, ambiguity related to costs and constraints, uncertainty about how systems really work, and many other factors make it unwise to accept the totally unwarranted level of precision that can be obtained from modern computational technologies.

My class on project evaluation was, like Caesar's Gaul, divided into main three parts. The first part provided an overview of project evaluation as a multi-dimensional process aimed at creating projects that meet the needs of society. The second part covered discounting, net present value, financial assessment, and other basic methodologies of engineering economics. The third part addressed issues such as risk and uncertainty, technology scanning, public-private partnerships, and the evolution of infrastructure systems over long periods of time.

Over time, the basic framework remained unchanged, but I was able to develop ever more detailed notes, additional assignments, more open-ended case studies, and more complete presentations for my undergraduate class on project evaluation. I also gave lectures on project evaluation in graduate courses at MIT in the Department Civil & Environmental Engineering, the Center for Transportation & Logistics, and the Engineering Systems Division. After retiring from my full-time appointment at MIT, I began to transform my lecture notes and other course materials into a series of essays and case studies suitable for a textbook. At the request of Jenny Welter, an editor at John Wiley & Sons, I expanded my notes by adding a great many simple examples, hundreds of problems, and new material on project management and engineering economics. In 2011, Wiley published *Toward More Sustainable Infrastructure: Project Evaluation for Planners and Engineers*, a 500-page textbook that covers the basic methods of project evaluation, provides examples attuned to infrastructure systems, and includes case studies that illustrate the breadth and excitement of project evaluation as related to infrastructure systems. Solutions to the problems, an instructor's manual, and power point presentations for each chapter are all available from Wiley. These materials can provide students and instructors with tools and concepts that they can use in understanding or teaching the need for projects, the options that are available, and the methods for evaluating and refining the options that are available.

However, a 500-page textbook is not the ideal format for presenting the concepts of project evaluation to a broader audience that includes grad students interested in infrastructure systems, mid-career engineers making the transition to management, public officials involved with infrastructure systems or anyone else with an interest in planning for, management of, or investment in infrastructure systems. I therefore decided to return to my class notes and professional papers in order to create a shorter, more focused book that would be readily available to anyone interested in infrastructure systems. Instead of a textbook with long chapters and hundreds of examples and problems, this book focuses on concepts and case studies directly related to project evaluation. It assumes the reader is familiar with supply & demand and other basic economic concepts; it does not cover project management; and it avoids going into esoteric elements of engineering economics such as equivalence relationships involving gradients or geometric sequences. Nevertheless, most of material in this book is very similar to what is in the textbook, because both books draw upon the same notes, case studies, technical papers, and presentations that I developed while teaching my class on project evaluation between 1997 and 2009.

The material includes two categories of documents, namely essays and case studies. Those who wish to gain a broad conceptual framework for understanding project evaluation in the context of infrastructure systems can read the essays; those who wish more detail on methodologies in the context of specific projects can concentrate on the case studies. Each essay and each case study is a stand-alone document that be read without being distracted by references to definitions or methods developed in prior or subsequent chapters. *Project Evaluation: Essays and Case Studies* should therefore be useful to practitioners and anyone with a general interest in project evaluation or infrastructure, even though it may be less appealing to a professor hoping to find a multitude of simple examples and a great many problems for his students to solve.

Although this book does not include sample problems and problem sets, such materials can be found under "1.011 Project Evaluation" as part of MIT's Open Courseware website at www.MIT.edu or directly from:

http://dspace.mit.edu/bitstream/handle/1721.1/75001/1-011-spring-2005/contents/index.htm?sequence=5

This web site provides the syllabus, reading lists, assignments, quizzes and other class materials for several different versions of the class. It also includes student presentations for a half dozen major projects, each of which would be interesting to a general reader of this book. The URL shown above is for the 2005 version of "Project Evaluation", which is the most complete version on Open Courseware for the years when I alone was responsible for this subject.

Structure of *Project Evaluation Essays and Case Studies*

Project Evaluation: Essays and Case Studies is published in two stand-along volumes. The first volume provides an overview of project evaluation as a multi-dimensional process aimed at creating projects that meet the needs of society. The essays and case studies in this volume provide a framework for understanding and evaluating projects, taking into account not only the financial and economic issues, but also social and environmental factors. The essays in this volume emphasize that analysis will not necessarily determine what projects are considered, what projects are proposed, what projects are approved or what projects are ultimately successful. Projects may be motivated by a vision of a greater society, by an idea for addressing a specific local problem, by the prospects of making a profit while providing a needed service, or by simple greed. Some apparently excellent projects cannot be financed, while it may be easy to fund some very questionable projects. Case studies in Volume I are mostly based upon actual infrastructure projects.

Volume II examines the equivalence relationships that can be used to compare cash flows or economic costs and benefits over the life of a project. It covers the concepts and methodologies that can be used by investors, bankers, and entrepreneurs in deciding whether or not to finance projects, and it shows how public policy can use taxes and other regulations to encourage projects that have public benefits. Most of the case studies in this volume present hypothetical situations that illustrate how various methodologies can be used in project evaluation.

Carl D. Martland
Senior Research Associate and Lecturer (Retired)
Department of Civil & Environmental Engineering
Massachusetts Institute of Technology
October 2016

Deer Island Sewage Treatment Plant:
Part of a $6 Billion Project that Helped Clean Up Boston Harbor.

Introduction

The first half of this book consists of six essays that describe a variety of methods and procedures that can be useful when evaluating potential infrastructure projects. The second half includes eleven hypothetical case studies that illustrate how these methods and procedures might be used. Each essay and each case study can be read as a stand-alone document, as there are no cross-references within them, nor is it necessary to read any essay before reading any case study. Someone familiar with the methodology may prefer to read about applications in their areas of interest. Someone with experience in particular kinds of applications may prefer to gain broader exposure to the methods described in the essays.

System Performance, the first essay introduces various measures that can be used to assess the performance of infrastructure projects and to evaluate alternatives for improving performance. Inevitably, there will be many aspects of performance to consider and many possible impacts on society or the environment that must be minimized or mitigated. Financial analysis, which is concerned with the cash flows directly related to a project, will be critical, but so will economic analysis, which also includes the impacts of a project on the overall economy. Both financial and economic impacts can be measured in monetary terms; which types of impacts are considered will depend upon who is doing the analysis. Owners, developers and users will largely be interested in financial matters; public agencies that must approve projects are concerned with broader economic matters, such as job creation and regional prosperity.

The next four essays are concerned with engineering economics, which provides many of the methodologies that are needed for financial and economic analysis, including the effects of taxes and depreciation, as illustrated in Figure 1. A central tenet of engineering economics is that it is possible to use a discount rate to estimate the equivalent present value of any future value. Given a discount rate, it is possible to calculate the net present value (NPV) of any stream of financial or economic costs and benefits that might be associated with a project.

Equivalence of Cash Flows develops the basic relationships that can be used to transform an arbitrary stream of financial or economic benefits into an equivalent present value, an equivalent value at some future time, or an equivalent annuity. These relationships are what makes it possible to compare the financial and economic impacts of multiple alternatives. The alternative with the highest net present value will also have the highest future value and produce the largest equivalent annuity. Thus, from a financial or economic perspective, it makes sense to choose the alternative that maximizes net present value.

The equivalence relationships all depend upon the choice of a discount rate, which is far from a simple, objective task. *Choosing a Discount Rate* goes into considerable detail discussing the factors that will affect the choice of a discount rate, emphasizing that different actors involved in implementing, using, or investing in a project may have different perceptions of the project and therefore may use different discount rates when evaluating a proposal.

Once the NPV of financial or economic benefits have been calculated, it is straightforward to select the alternative with the highest positive NPV. However, complications may arise if a different measure is used. Companies commonly use the internal rate of return when evaluating projects. The IRR is the discount rate that makes the NPV equal zero. If the IRR is greater than a company's discount rate, then that is a valid project. However, if there are mutually exclusive projects, then a small project with a high IRR might appear to be better than a larger project with a lower IRR that actually has a higher NPV. *Financial Assessment* presents methods that show how to deal with this issue. So long as the IRR analysis is applied properly, it will provide the correct ranking of mutually exclusive projects.

Figure 1: Structure of Four Essays Related to Discounting and Cash Flows

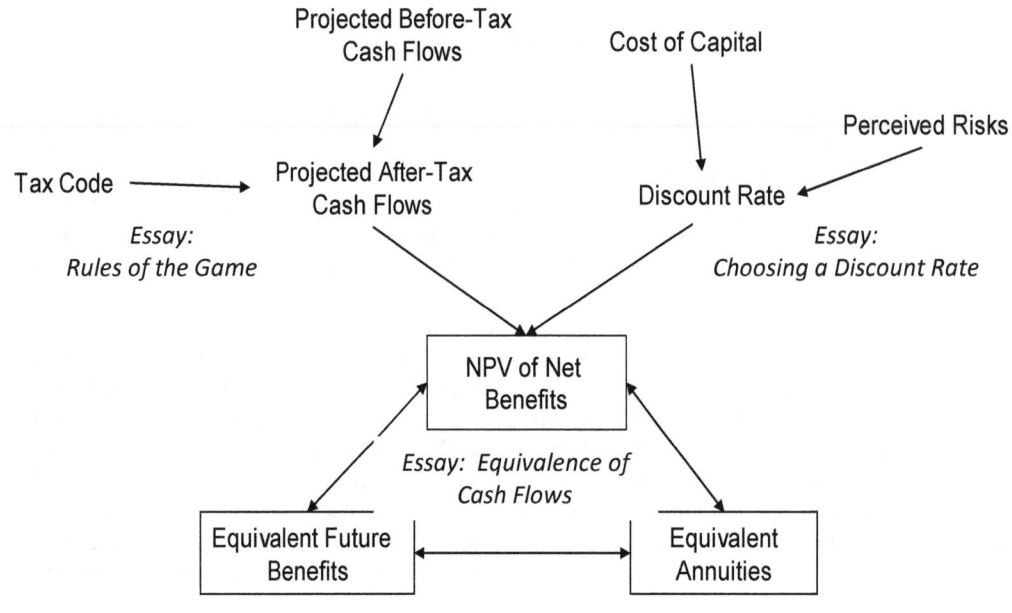

Public policy can affect the cash flows of a project in many ways. Zoning restrictions may limit what can and cannot be done on a site. Regulations may determine what kinds of materials or construction techniques can be used. The intricacies of the tax code can be manipulated by law-makers in order to promote or hinder certain types of development or certain types of investments. In particular, major investment expenses usually cannot be immediately deducted from taxable income; instead capital investments are depreciated over many years. Exactly how depreciation is treated in the tax code will determine when this expense is incurred. Since expenses affect profits, and profits result in income tax payments, it is necessary to consider depreciation in order to obtain a valid after-tax analysis of a project. *Rules of the Game* shows how depreciation, taxation, zoning, and environmental regulation can evaluation.

Dealing with Risks and Uncertainties introduces methods that are commonly used in project evaluation to deal with risks and uncertainties: modeling performance, probabilistic risk assessment, and performance-based technology assessment (PBTS). One case study uses probabilistic risk assessment to examine ways to reduce the risks associated with rail-highway grade crossings. Another case study uses PBTS to examine competition between airlines and railways for intercity passenger traffic.

System Performance

As it is always easier and in the end less costly to be accurate than inaccurate, the good engineer will always be accurate in all essentials, but he will not waste time in attempting unnecessary precision which does not add appreciably to the final value of his work.

Arthur M. Wellington, **The Economic Location of Railways,** John Wiley & Sons, 1911

Performance of Infrastructure-Based Systems

The performance of an infrastructure system cannot be captured by just one or two measures. Cost, quality of service, capacity, safety, environmental impacts and sustainability are all important, while the extent of coverage, accessibility, equity, and appearance can also be critical. Moreover, performance depends upon one's perspective, as owners and managers seek financial rewards, users seek good service at reasonable prices, and the public worries about such things as the need for subsidies, environmental impacts, safety, land use, economic development, and aesthetics (Figure 1). Designing, constructing, and managing these systems involves trade-offs among multiple factors, often with no obvious way of determining which factors are most important. Evaluating proposals for creating new systems or for expanding or modifying existing systems will always require judgment and will often require some sort of political process to determine which aspects of performance should be emphasized.

System performance can be documented, studied, modeled, predicted and managed. If system performance is well understood, and if there is a consensus about the relative importance of the various aspects of performance, then it is possible to provide a clear, objective basis for evaluating new projects and programs. If performance is poorly understood, then research and analysis may be able to clarify the potential trade-offs for various options. If there is no consensus as to which aspects of performance are most important, then research and analysis of performance can at least provide an objective framework for evaluating proposals.

Figure 1 Perspectives on Infrastructure Performance

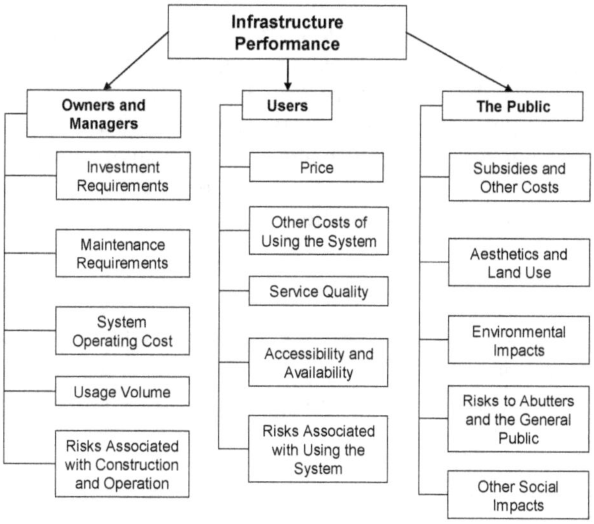

Performance of infrastructure systems depends upon engineering and managerial issues, such as the nature, condition, and deterioration rates of structures and equipment. Infrastructure managers develop plans and policies that they use to guide operations and maintenance, usually with the hope of increasing profitability or other financial goals. The management structure will often have separate departments responsible for operations, maintenance and marketing, each of which will have their own concerns and ideas about what types of projects will be most beneficial in improving performance. Each department's objectives will ideally reflect strategic plans, which may include financial goals, goals for expanding or shrinking the system, plans for new services or markets, or goals related to service quality, risk management or interactions with the public (Figure 2).

Figure 2 Managing Infrastructure

Performance also depends upon many other factors, including the regulatory structure, operating capabilities, and business strategies pursued by the organizations and individuals that make use of the infrastructure and the existence and performance of competing infrastructure systems.

Most importantly, perhaps, performance reflects the demand for the service, especially when usage approaches system capacity. A highway that allows motorists to drive through the city at high speeds in the middle of the night can be more like a parking lot during rush hour. It is a very complex matter to predict the average speed on a highway during rush hour, for that will require some way to estimate demand taking into account the fact that the level of demand will affect performance. It is more straightforward to develop a **performance function** for the highway, i.e. to predict the performance of the highway for any particular level of demand.

Performance functions can be developed for any infrastructure system based upon the engineering characteristics of the infrastructure and the ways that it is used. Developing **engineering-based performance functions** enables planners and entrepreneurs to understand the options for developing or improving systems, to evaluate the impact of proposed projects, to understand the potential benefits of new technologies, and to select and design better projects. Performance functions may or may not be precise, depending upon the state of knowledge of the system and also upon the context for their use. The remaining sections of this essay focus on performance measures important for all infrastructure systems: cost, profitability, service, capacity, safety, and security.

System Cost

Owners and investors are concerned with the initial cost to construct infrastructure, as well as the continuing cost for operating and maintaining the infrastructure. Users are concerned with their costs of using the system, which would include the prices they are charged plus their own time and expense associated with using the system. Basic cost concepts include the following:

- Total cost: what is the total cost of the system for a given level of output?
- Average cost: what is the average cost per unit of output?
- Marginal cost: what is the cost for one additional unit of output?

- Incremental cost: what is the cost for an increment of output?

These concepts can be illustrated using an extremely simple equation expressing cost as a linear function of volume V:

(Eq. 1)   Cost = a + b V

(Eq. 2)   Average cost = (a + b V)/V = a/V + b

(Eq. 3)   Marginal cost = (a + b (V + 1)) – (a + b V) = b

The fixed cost "a" is incurred whatever the volume, and "b" is the marginal cost per additional unit. The average cost per unit declines as the fixed cost is spread over additional volume. In any particular situation, these cost functions will be much more complex. Total cost will be a function of the resources that are used and their unit costs, and the resources that are used will depend upon the quality of service as well as the level of demand.

The fixed cost could well be the sum of many different types of costs that would not vary with changes in volume, such as the salaries of the senior administrative staff, the costs of establishing a maintenance facility, and the costs of acquiring basic equipment and machinery. Variable costs would likely include the labor and energy costs associated with operations. Which costs are fixed and which costs are variable depends upon the level of volume and the time frame under consideration. In the short run, many more costs are fixed, while in the long run, most costs can be affected by restructuring the system. Whatever the time frame, the simple cost function shown above as Equation 1 may (for a reasonable and meaningful range of volumes) actually be a useful approximation of a much more complex cost function. Hence, it is worth taking a closer look at this cost function.

Consider a situation in which following cost function is believed to be approximately valid for volumes V ranging from 10 to 100 units per day:

(Eq. 4)   Total Cost = $50 + $1 (V)

The fixed cost is $50, which plots as a straight horizontal line in Figure 3, while the variable cost plots as a straight line with a slope of 1. If this line were extended beyond the range of interest, it would intersect at the origin of the graph.

Figure 3 Plotting fixed, variable and total cost if C = 50 + V

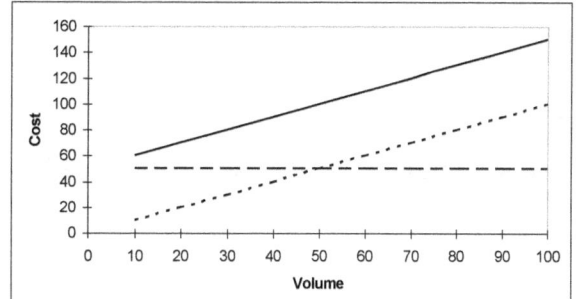

Figure 4 shows the average and variable cost for this cost function. The marginal cost is equal to $1 over the entire range, while the average cost declines from $6 when volume is 10 units per day to $1.50 when demand is 100 units per day. The average cost function is non-linear, and the average cost approaches the marginal cost for high volumes.

Figure 4 Plotting average and marginal cost if C = 50 + V

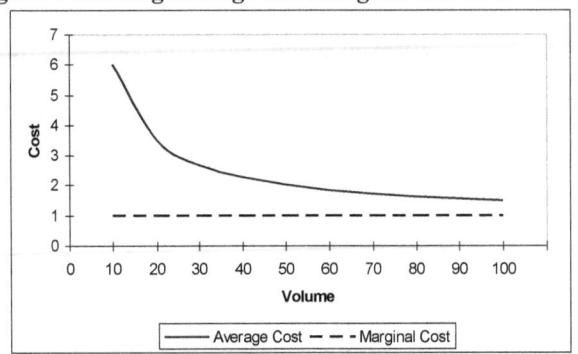

It is very common to find that there is a trade-off among two options, one of which has higher fixed cost but lower marginal cost. Figure 5 shows a second cost curve in which the fixed cost has increased from $50 to $95 per day, but the variable cost has dropped from $1 to $0.50. The breakeven point is at 90 units per day: above this level, option 2 is preferred, while below this level option 1 is preferred.

**Figure 5 The option with the higher fixed cost is preferred if the volume is greater than 90
(Cost of Option 1 = 50 + V; Cost of Option 2 = 95 + V/2)**

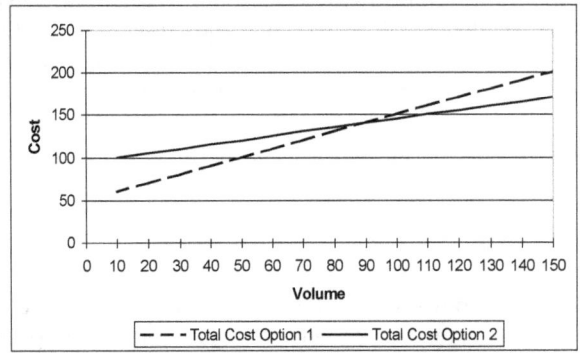

For linear functions of the type TC = a + bV, the breakeven point V_b can readily be calculated:

If $\quad\quad\quad TC_1 = a_1 + b_1 V$

and $\quad\quad\quad TC_2 = a_2 + b_2 V$

then, at the point where the costs are equal for the two technologies, the following equation will hold:

(Eq. 5) $\quad\quad a_1 + b_1 V_b = a_2 + b_2 V_b$

Solving for V_b yields:

(Eq. 6) $\quad\quad V_b = (a_2 - a_1)/(b_1 - b_2)$

In words, the breakeven volume is calculated as the increase in fixed cost divided by the savings in variable cost per unit. Major CEE projects require investments that are aimed at reducing marginal cost. Larger projects typically provide an opportunity for greater reductions in marginal cost. A key question is whether there will be enough demand to cover the extra costs associated with the larger project. In general, smaller projects will be better until congestion and capacity concerns make a larger project desirable.

> L.F. Loree, Past-President of the Delaware & Hudson Railway and Chair of the Kansas City Southern, described how railroads worked hard to avoid investing unnecessarily in capacity:
>
> *We hang back and postpone as long as possible work to increase facilities the use of which may be increased by ingenuity and method.*
>
> L.F. Loree, **Railroad Freight Transportation**,
> D. Appleton & Co., 1922, p. 61

Non-linear Cost Functions

Linear cost functions are easy to draw, but more complex functions will often be necessary. The same logic applies to plotting total costs, average cost, or marginal costs and the same approach can be used to determine breakeven volumes. Figure 6 illustrates a situation where three technological options are available, each of which can be represented by a non-linear cost function:

(Eq. 7)   Total Cost Option 1 = $50 + V + 0.03V^2$

(Eq. 8)   Total Cost Option 2 = $100 + 0.5V + 0.02V^2$

(Eq. 9)   Total Cost Option 3 = $300 + 0.25V + 0.01V^2$

The third term in these equations means that the average costs eventually begin to increase, as the contribution of the squared term eventually offsets the savings from spreading the fixed costs over a larger volume. Figure 6 plots the average costs of the three technologies for the ranges of volumes for which they are feasible options. Option 1, the one with the lowest initial cost, has average costs greater than $6/unit for the minimum volume of 10 units; average costs drop below $4/unit for volumes of 30 to 50 units, then rise steadily as volume increases to 100 units, which is the maximum that can be handled by this technology. Option 3 is much too expensive for or ill-suited to low volumes, so the costs for that technology are only shown for volumes of at least 40 units. Figure 6 indicates that Option 1 is favored for volumes less than about 50, while Option 2 is cheapest for volumes between 50 and 130 and Option 3 is favored for greater volumes.

Figure 6 Long-Run Average Costs

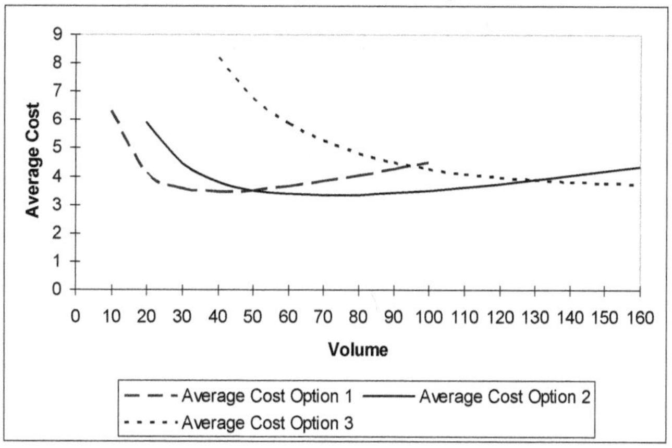

Short-Run vs. Long-Run Average Costs

The three cost curves shown above in Figure 6 each depict **short-run average costs** for a particular technology. Each curve shows the average costs that would result from using a particular technology for a specified range of volumes. They are called "short-run" average costs because they do not allow a shift to a more efficient technology.

It is often useful to understand how average cost of a particular type of process or production would change with volume, assuming that the best technology is used for each level of volume. This cost is called the **long-run average cost**, assuming that in the long-run it will be possible to adjust the production process to what is best for the actual volume served or produced. The long-run average cost curve can be constructed from the applicable short-run cost curves: for each volume, the long-run average cost is equal to the short-run cost of the technological option with the lowest short-run average cost. In other words, the long-run average cost will be the envelope of the minimum short-run costs, as depicted in Figure 7. This is a single cost function that is based upon the three cost functions shown in the previous figure. The long-run average cost in Figure 7 could be described as a little less than $4 for volumes ranging from 20 to 160.

Figure 7 Long-Run Average Costs for Technologies Shown in Figure 2-6

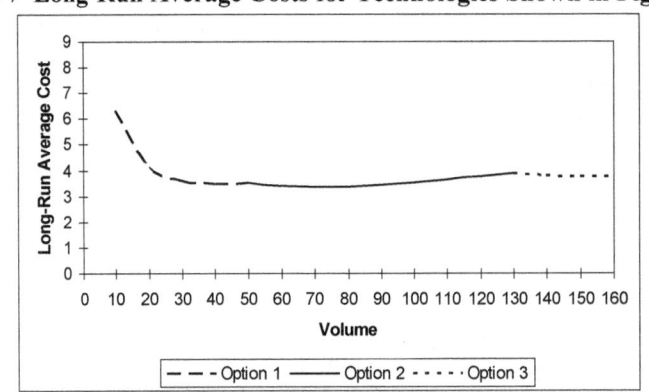

You (the developer, entrepreneur, or planner) know, or should know, your costs and technologies. You therefore should be able to develop an algebraic expression for your costs that accounts for the technological and design options that you may have. It the developer, entrepreneur, or planner doesn't know their costs or technologies, then you (the consultant, the researcher, or the smart young analyst) may be able to do some analysis and create relevant cost functions.

> *There is more than one way to skin a cat.*
> American Proverb

Resource Requirements

Costs ultimately are linked to resources: people, land, materials, energy, and capital. There will always be many different ways to use resources in constructing a particular project, and choices will probably be made by developers, engineers and planners seeking to minimize the cost of the project. From their perspective, minimizing cost is purely a financial exercise aimed at minimizing the cash required to complete the project.

The design of a project and the resources required for it will be influenced by both the current and the expected future availability and prices of people, land, materials, energy sources, and capital. If average salaries and wages are expected to rise relative to the cost of capital, then there will be a tendency to use fewer people and more machines. If the cost of energy is expected to rise relative to the cost of building materials, then there will be an incentive to construct buildings that require less energy and to use materials that are less energy intensive. If longer lasting materials are created, they will be used where their longevity provides benefits to a project. Standard methodologies of engineering economics make it possible to decide what system design is best for a project, which materials to use, and which construction techniques to follow, taking into account the initial investment cost and costs to the owners and users over the life of the project.

From an economic perspective, however, the price of a resource may not reflect the true economic cost of using that resource:

- The cost of using a resource is an **opportunity cost**; the use of resources on one project means that they are not being used on other projects or being reserved for future use.
- There may be **negative environmental impacts** associated with using a resource, such as the paving over of wetlands, emission of greenhouse gases in the manufacture of construction materials or the devastation of remote wild areas by mining activities.
- **Non-sustainable use** of renewable resources, such as ground water or timber, will lead to declines in the amount of such resources that can be used in the future.
- Extensive use of non-renewable resources can eventually lead to **resource depletion**.
- Exposure of workers to unsafe conditions or hazardous environments may lead to **risks of injury or serious illness**; unscrupulous companies may be able to avoid covering the costs associated with these problems, and governments may or may not regulate workplace conditions.
- There may be **government subsidies** for certain types of workers or government regulations that require the use of excessive numbers of workers.

Some of these costs may be reflected in the market prices for certain resources, such as the cost of capital and salaries and wages in developed countries. Other costs are notably absent from the prices that are charged, such as the effects of emissions on global warming or the environmental damages related to strip mining. If a company or a country already owns land that could be used for a project, they may not even consider the cost of using that land in the analysis, even though developing that land may forestall even better opportunities in the future.

If the prices for resources actually approximate the costs associated with using those resources, then financial analysis will perhaps produce a reasonable result. If the prices for some resources are markedly above or below their full economic costs, then the financial analysis could result in poor decisions from the perspective of society, even when those decisions do appear to increase the profitability of the project. Because of the discrepancy between financial and economic costs, public policy must enforce rules and regulations that promote or enforce consideration of the true economic costs of a project. Such regulations include:

- Restrictions on development of wetlands and other sensitive environments
- Regulations related to occupational safety and health
- Minimum wages and laws allowing the formation of unions
- Licensing of engineers and others involved in project design and implementation
- Regulations concerning the technologies and methods used in mining and other extractive activities
- Regulations requiring assessment of environmental and social impacts

For most privately sponsored projects in developed countries, project evaluation will be strictly based upon a financial analysis. If government regulations are effective, then the private decisions will be reasonably good. If government regulations fail to capture significant aspects of economic costs – as has been the case with many environmental impacts – then private decisions could lead to increasing problems for society.

When evaluating projects, or when estimating costs, it will always be worthwhile a) to detail the actual resources that will be used in addition to how much money is needed and b) to document the methodologies and logic used in reaching decisions about design, choice of materials, and choice of construction methods. As prices of resources change relative to each other, as new technologies become available, and as more economic aspects of costs are recognized, different types of projects and different approaches to constructing those project will be needed. However, the original methodologies and logic may still be suitable for evaluating future projects.

Since many engineers and planners work in places with markedly different levels of development, it is important to recognize that the best solutions in one country can be quite different from the best solutions in another country. This is especially true when comparing systems in a developed country with systems in a developing country, where labor costs are apt to be much lower and capital costs much higher.

Lifecycle Cost

Since infrastructure lasts a long time, infrastructure design should consider costs over the entire life of the infrastructure. As shown in Figure 8, initial costs of design are likely to be modest compared to the construction costs that follow. Once the project is completed, there will costs of operation that may be borne by the owner, users or abutters. As the facility ages, it may be desirable to expand or to rehabilitate the facilities. Eventually, as a result of declining demand or excessive costs of maintenance, it will be necessary to decommission the facilities and to salvage metal or whatever is left of value. Notice that the original owner and developer incur the initial costs, while users and abutters and subsequent owners are left with whatever the long-term costs turn out to be.

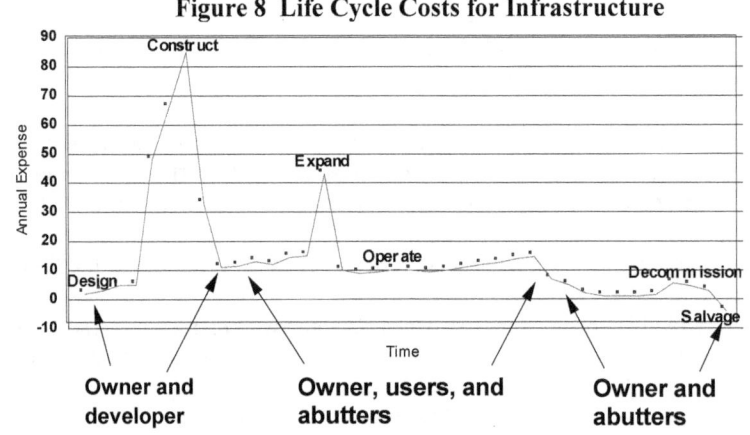

Figure 8 Life Cycle Costs for Infrastructure

The design phase offers the greatest opportunity to affect the life cycle costs of a project (Figure 9). At this point, while all things will not be possible, many things will be. A major design consideration will be the extent to which the owner or developer considers the costs to users and abutters. Small changes in design conceivably produce more efficient operations or limit the negative effects on third parties, but only to the extent to which such costs are even considered in the design. Mistakes regarding the size of the project – too big, too small, too soon, too little flexibility, too difficult to rehabilitate, too much impact on neighbors – conceivably can be rectified with little or no additional time or expense related to construction. The opportunities for savings will be clearest when the owner will be responsible for operations for the indefinite future; the opportunities for misguided design (or fraud) will be greatest when the developer is interested only in minimizing the construction cost, as operations and maintenance will be the responsibility of others.

Figure 9 Opportunities for Reducing Cost Are Greatest at the Outset

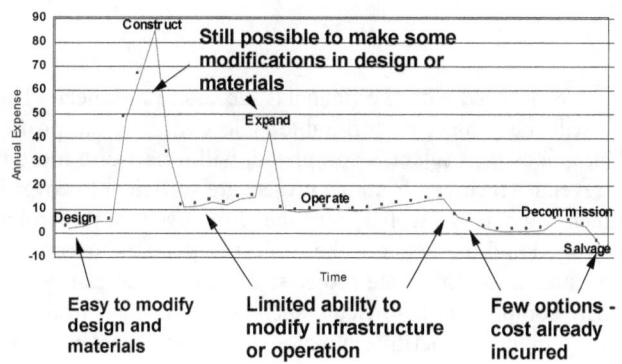

Engineering-Based Cost Functions

Three approaches can be used to estimate cost functions: accounting, econometric, and engineering-based. Each approach has its usefulness, but for project evaluation, an engineering-based approach is essential.

Accounting approach: for an existing system, whether an apartment building, a transportation terminal or a wind farm, it should be possible to identify all of the expenses related to the system's construction and operation. If the accounting system is accurate and complete, then the costs associated with each phase of the design, construction and operation of the project should be evident. Complications will arise if there are many different owners and users or if it is difficult to allocate specific costs to specific purposes or users. Special studies may be required to support a realistic allocation of costs or to identify user costs. However, the companies or agencies that manage infrastructure are likely to have very good information concerning their own costs and good estimates of the costs borne by their users.

There are two main problems with using accounting costs for project evaluation. First, accounting costs will not readily relate to the costs of new projects, unless the new projects are very similar to existing projects. Second, accounting systems can provide very rich detail concerning existing operations, but they will not show how costs vary with demand, quality of service, capacity or technology.

Econometric approach: if cost data are available for a variety of completed projects, it may be possible to discern trends in cost as a function of the type or size or location of the project. Econometric analysis can be useful for public policy. For example, if data were available on the cost of constructing apartment buildings in a major city, it would be possible to determine the cost per apartment or the cost/square foot of living space for each project. This data could then be plotted to determine if the cost/apartment and the cost/square foot vary with the size of the project. If large projects involving a hundred or more apartments are markedly cheaper to construct than small projects, then public policy perhaps should be slanted toward facilitating larger projects.

Although the econometric approach can be very useful in determining some basic trends in cost, it is less useful in estimating costs for a particular project. It is very ill-suited toward estimating the costs of projects that use new designs or technologies – and these of course are among the characteristics of many large-scale infrastructure projects.

Engineering-based approach: the preferred method for estimating infrastructure costs is to break a project into well-defined pieces for which it is feasible to estimate unit costs. For each piece of the project, unit costs can be developed based upon past experience, expert judgment, or special studies. These unit costs can reflect new technologies or new designs, so that this approach is not restricted to past experience.

Profitability, Breakeven Volume, and Return on Investment

Profit

Entrepreneurs and owners will be concerned with the financial success of a project. There are three main questions that will be of interest. First, will the project be profitable? Second, will the profit be sufficient to justify the investment that is required? Third, once the project is completed, will it be worth more than it cost to build it? The project will be profitable if the revenues received from the project are sufficient to cover its costs. The revenue could include subsidies from government agencies as well as revenues from users of the project. Revenue from users will depend upon the price that is charged and the value of the project to potential users. For projects that add capacity within a competitive market, such as most real estate projects, the prices that can be charged will rise and fall with market forces. For projects where competition is difficult, such as new bridges or toll roads, the prices can to some extent be established by the owner. In the competitive situation, the question is whether the project can be constructed and operated so that it is possible to achieve a profit given expected market prices. In a monopolistic situation, the question is to choose the prices that will maximize profits, assuming that there is in fact a range of prices that could be profitable.

Let's begin with the simple situation we discussed in the previous section:

(Eq. 10)   Total cost = a + bV

If this is a project that will add capacity to a competitive market, then the price P will be determined by the market and the total revenue can be expressed as:

(Eq. 11)   Revenue = PV

Profit will be the difference between revenue and cost:

(Eq. 12))   Profit = PV – (a + bV)

Breakeven Volume

If a company has a linear cost function of the type shown above in Figures 1 and 2, then it must sell enough units so that the average cost of production is less than the average sale price. If the sale price is less than the variable cost, then the company will never make a profit. If the sale price is higher than the variable cost, then the company will receive enough cash from the sale to cover the variable cost and have something left over that could go toward covering fixed costs; once sales volume is sufficient to cover fixed costs, then each sale will add to profit. The difference between the sale price and variable cost is called the **contribution to fixed costs or profit** or the **contribution to overhead or profit**:

If   Cost = a + bV

and   Price = P

Then

(Eq. 13)   Contribution = P – b

(Eq. 14)   Profit = PV – (a + bV) = (P-b)V – a

Sales will be enough to generate a profit if the total contribution is enough to cover the fixed costs. The point at which contribution equals fixed cost is known as the breakeven volume V_p relative to making a profit:

(Eq. 15)   V$_p$ = a/(P – b)

There are two conditions for profitable operation in this simple situation: price must be above variable cost and the volume must be above the breakeven volume. V$_p$ is the volume where the incremental contribution to profit from each unit produced is sufficient to cover its share of the fixed cost.

Volume Varies with Price – the Demand Curve

More generally, usage volume will vary with the price that is charged. The higher the price, the lower the volume. This is traditionally represented as a demand function, as shown in Figure 10. Note that price, presumably the independent variable, is shown on the y-axis while volume, presumably the dependent variable, is shown on the x-axis. Faced with a downward sloping demand function, someone trying to decide upon a price for their services is faced with a dilemma. Set the price too high and no one will show up; set the price too low and there won't be any profit, no matter how high the volume. The expression for profitability will be the same as shown above (Eq. 14), and it will still be necessary for price to exceed variable cost to make a profit.

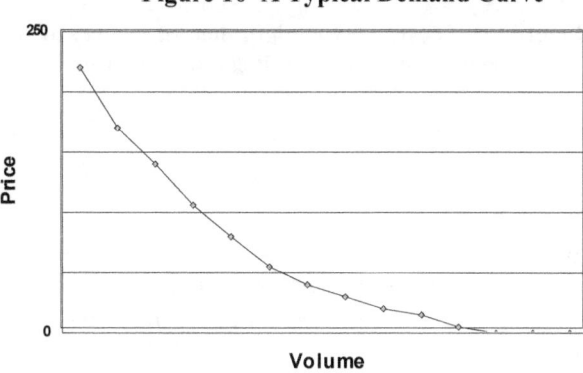

Figure 10 A Typical Demand Curve

The best price – at least from the owner's purely financial perspective - will be the price that maximizes profit. Figure 11 plots profit as a function of volume, where the price is defined by the demand curve shown above. The first point on each curve corresponds to a price of $225 per unit and a volume of 1 unit per day. The third point corresponds to a price of $140 and revenue of $420 for sales of 3 units per day; this is the price that maximizes profit. If price is further lowered to $110, the volume increases to 4 units per day and revenue rises to $440 per day, but profit declines. With further price reductions, the total revenue declines. When price drops to $15, the volume is much higher, but total revenue equals total cost. For even lower prices, the owner loses money. The project therefore would be profitable for any price between $15 and $225/unit. Note that the owner would rationally choose to keep prices high, even if there were additional benefits to society to be gained by attracting more users.

There will not necessarily be any price at which the owner could make a profit, as illustrated by Figure 12. If total cost always exceeds total revenue, then private companies will not provide the service. If the service is deemed to be desirable for society, then the public may decide either to provide the service via a public agency (e.g. public transit) or to subsidize the service so that the private sector will continue in the business (e.g. subsidized housing for low-income families).

Figure 11 Maximizing profit is not the same as maximizing revenue.
As prices fall, volumes rise, but total revenue eventually drops below total cost.

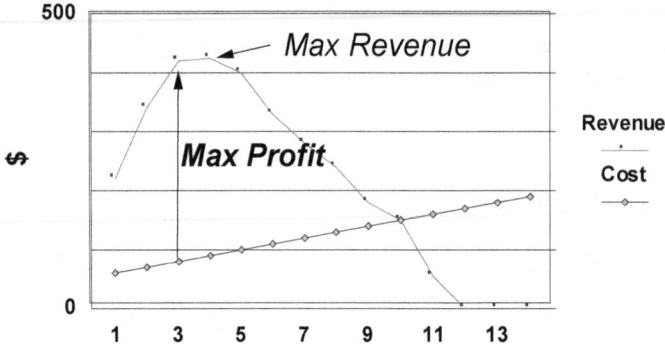

Prices people are willing to pay for use of infrastructure, for renting space in an office building, or for any other service or product will depend upon many factors out of control of the company or agency that is trying to make a profit. General economic conditions and the overall supply of office space sometimes lead to dramatic changes in building occupancy and rental rates.

Figure 12 If costs are too high, then there may be no price for which a profit can be achieved

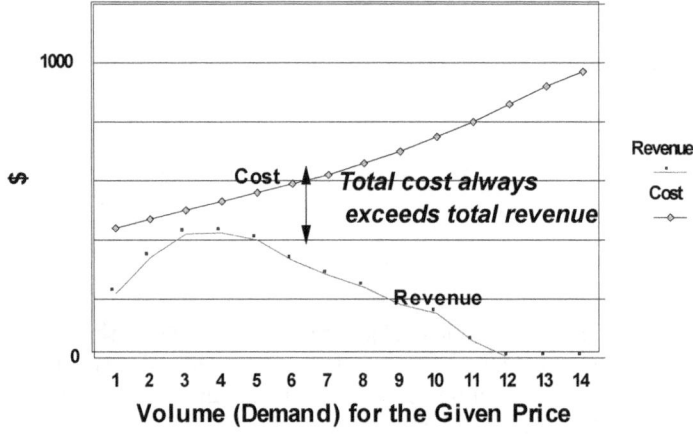

Return on Investment

The financial success of a project is often measured by the return on investment, which is the annual profit divided by the initial investment in a project. If someone invests $100,000 in a project that provides a profit of $15,000 per year, then the return on investment is $15,000/ $100,000 or 15%. Usually the return on the investment will vary from year to year, because prices and usage volume will vary with the economy, competitive conditions, and other factors that will affect supply and demand. If return on investment is high, then investors will consider additional investments in similar projects; if returns are low or negative, then investors will turn to other types of projects, leave their money in the bank, or perhaps invest in other companies by buying their stocks and bonds.

Service

Potential customers for a project will be concerned about the quality of service as well as the cost. Moreover, there may be many different aspects of service quality that could be important to certain customers. For example, average trip time, trip time reliability, the probability of excessive delays, comfort, and convenience can all be important for passenger transportation systems. The relative importance of these measures will vary with the type of trip, the type of traveler, and the travel options that are available. For the movement of freight, key measures include the size of the shipment that can be carried and the probability that the shipment will be lost or damaged in addition to trip times and reliability. During the nineteenth century, the introduction of railroads revolutionized inter-city travel as it was so much faster than walking, riding a horse, or going down a canal in a barge. During the 20th century, the introduction of automobiles and air travel drastically reduced the role of railroads and led to massive investments in infrastructure for highways and airports. During the 21st century, high oil prices, congestion, and concerns about emissions may result in another major realignment of transportation systems and investments in transportation infrastructure.

Water resource systems have different measures of service quality. In developed countries, the main concern is likely to be the price, quality and reliability of the water supply. Potable water will be needed for every household, and sufficient quantities of water will need to be provided for all uses and all users at a reasonable price. In many regions of the world, where running water is a luxury, other measures will be useful, such as the percentage of homes with running water, the average distance to the nearest clean water in rural villages, and the probability of disruptions in supply. In some regions, the chemical content of the water is a major concern for public health. For example, arsenic has been found in rural water supplies in Nepal, resulting in serious health problems in many villages.

Historically, failure to keep drinking water separate from waste water and sewage resulted in outbreaks of cholera and other diseases that periodically would kill tens of thousands of people in major cities. Contamination of drinking water remains a critical problem today in situations where earthquakes or tsunamis endanger water supplies and in regions where wars or rumors of war cause thousands of people to move to refugee camps with inadequate infrastructure for either water or sewage.

Navigation, flood control and the use of water power to generate electricity are other matters that relate to the construction of dams and levees. Riverbanks and lakeshores are wonderful locations for parks – but also prime spots for industrial uses. The conflict between the natural and the manmade environment is very apparent as wetlands are filled in, whether to provide additional space for ports, seaside housing, or industrial uses. In many rural areas, vast expanses of wetlands have been drained (via extensive networks of drainage pipes, ditches, and channels) in order to provide fertile lands for agriculture.

Still other types of measures will be needed when considering energy projects. For example, the use of electricity is pervasive throughout the developed world. Normally, customers simply have access to electricity simply by flipping a switch, and their concerns about service relate to the time that the electric company requires to respond to infrequent outages caused by bad weather. Difficulties in finding an electrician to install a new appliance or disputes with the power company about billing are likely to cause more concerns than the availability of power. In the rest of the world, the availability of electricity can be a dominant concern, as many regions have either no access to the power grid or access that is limited by time of day or day of week because of limitations in capacity. Poorly maintained systems may have very frequent outages, and some systems are prone to disruption related to civil unrest.

Access to energy, limitations on the use of energy, likelihood of planned and unplanned disruptions, and the time required to restore service will be aspects of service related to other types of energy used directly by consumers, including gasoline, diesel fuel, natural gas, firewood.

Energy companies have options in terms of the kind of energy resources that they use and the origins of those resources. From their perspective, a critical measure of service quality is the reliability of the supply chain, which is the process by which coal, oil or other energy resource is mined, processed, and transported to their facility. In order to meet their customers' needs, energy companies prefer reliable supply chains for their fuel. If the supply chains are unreliable, companies have several options. They could maintain large inventories of fuel that will enable them to continue operations despite periodic breaks in deliveries, they could have multiple sources of fuel, or they could have redundant networks for providing their services, so that failure in one portion of the system does not shut down the entire system.

Engineering-Based Service Functions

As was the case with cost, engineering-based functions can be developed for each aspect of service. However, service is more complex than cost, because it is not possible to have a single measure (such as "dollars") that is readily accepted by owners and users and everyone else as a valid measure of performance. Thus there is a multi-step process in developing service functions. First, and probably most important, it is important to understand what aspects of service are most important to potential customers. Second, it is necessary to understand how design relates to the quality of service that can be offered.

Capacity

Capacity is another important aspect of the performance of infrastructure-based systems. Measurement of capacity is complicated by a number of factors, not the least of which is the difficulty in defining output. Infrastructure-based systems typically consist of networks of facilities that are managed or used by multiple organizations using labor, equipment, energy, and other resources to provide what may be a rather large array of multi-dimensional services. To understand capacity, it is best to start with a small piece of the system, where it is possible to enumerate all of the key inputs, where volume is well-defined, and where it is straightforward to estimate what happens to performance as volume increases. As with both cost and service, it is very useful to develop engineering-based functions that relate capacity to the characteristics of the infrastructure and its users.

It is often useful to consider three types of capacity:

- **Maximum capacity** – the maximum flow through the system when everything is operating properly
- **Operating capacity** – the average flow through the system under normal operating conditions
- **Sustainable capacity** – the maximum flow through the system that allows sufficient time for maintenance and recovery from accidents or other incidents

Since demand may vary substantially on a daily, weekly or seasonal basis, it is also important to consider how well the system performs during periods of peak demand. There could well be delays to users or denials of service during peak periods, but the system may be able to recover soon after peak demands subside. Delays during peak periods do not mean that capacity has been reached; it is only when delays become serious impediments to users that capacity has been reached. The solution to peak period problems could be expansions to the infrastructure, changes in operations, changes in pricing, or other attempts to limit peak period demands.

Safety and Security

Safety and security are important aspects of performance for infrastructure systems. Both refer to the likelihood of injuries or fatalities to employees, customers, abutters, and others along with the probability of damages to the infrastructure or to the property of users or abutters. Security generally refers to the measures that might be taken to prevent deliberate attacks on people or property, whether those attacks involve pickpockets, thieves, or terrorists. Safety is generally used in reference to accidents or problems that occur in the course of normal operations of the system, although it may also be used with respect to the possibility of deliberate attacks. Safety records for a transportation company show such things as the number of accidents, the number of fatalities, and the number of fatalities per million miles traveled.

Risk is a broader concept that can be used to describe the potential for future accidents or incidents.[1] A methodology known as "probabilistic risk assessment" can be used to measure risk in a way that is very useful in evaluating system performance. Risk is the product of two factors:

- The probability of an accident or incident
- The expected consequences if an accident or incident occurs

Consequences could include fatalities, injuries, disruption of service, release of toxic chemicals, or inconvenience to people living in or moving through the neighborhood of the accident. A weighting scheme will be needed to compare the severity of the different types of accidents. Strategies for reducing risks could focus either on reducing the probability of accidents or on reducing the expected consequences if there is an accident.

The design of infrastructure systems always involves consideration of risks, including risks to those involved in construction of the facility as well as risks to users. Construction standards and choice of materials will depend in part upon the potential for natural disasters, such as hurricanes or earthquakes, and the need to provide a safe operating environment. It is never possible to eliminate all risks, and judgment will be necessary to determine which risks are worth worrying about.

Some adjustments in predicted risks may be necessary to reflect the perceived importance of certain types of accidents or consequences to various stakeholders. Quantifying perceived risks requires answers to questions such as "Who is at fault?", "Is it a catastrophic accident?", and "Is new technology involved". Public perceptions of risk will be greater for situations where that risk could result in catastrophic accidents with hundreds of fatalities or accidents with dreadful consequences, as could be the case with the release of toxic chemicals or radiation after an accident at a chemical factory or at a nuclear power plant. The public is more concerned with unknown risks that might be associated with new technologies such as drones or driverless cars than with well-known risks. People know that automobile accidents kill tens of thousands of people per year, but they still drive cars. Fears of accidents involving nuclear power plants hampered development of such projects in many countries, despite an excellent safety record.

A general approach to investigating the risks related to infrastructure encompasses the following basic concepts:

- Risk reflects both the probability of an accident and the consequences of the accident.
- Past experience can be a guide for estimating accident probabilities and expected consequences.
- Weights can be devised for comparing different types of consequences.
- Strategies for reducing risks can be evaluated by comparing the costs of those strategies to the expected reduction in risk.

Companies, operating agencies and agencies that create, use or regulate infrastructure are continually seeking cost-effective ways to reduce risks. Since resources are limited and risks can never be entirely eliminated, it is important to allocate those resources effectively. In practice, this means that the expected reduction in the consequences of accidents should be high enough to justify the costs.

Cost Effectiveness

When dealing with costs and finances, it is natural to measure everything in terms of money. When dealing with the other matters, it is more difficult to use money as a performance measure. Thus, if an investment is proposed to improve infrastructure performance, it will usually be difficult if not impossible to determine whether or not the non-

---

[1] The term "risk" is used in two different ways by safety engineers and financiers. To safety engineers, risk refers to accidents or incidents that result in property damage, injuries or fatalities, and that is the sense in which the term is used in this section. In finance, the term is used to include any of the many factors that may affect the success of a project, including such things as the risk of an economic downturn, the risk of new competition, and the risk of political upheaval as well as the risks related to safety and security. Both meanings are used in this text; which sense is being used should be clear from the context.

monetary benefits justify the financial cost of the investment. Ultimately, some kind of political process will be necessary to weigh all of the costs and benefits to determine whether or not a proposed project is justified.

The concept of **cost-effectiveness** is a valuable technique that can be used in evaluating potential investments aimed at achieving non-monetary benefits. Cost-effectiveness is the ratio of the benefit to the cost of achieving that benefit. Thus, as long as it is possible to measure a benefit, it is possible to calculate cost-effectiveness. If there are multiple ways of achieving the same type of benefit, then the most cost-effective approach is the one that costs the least per unit of improvement. Just because an option is the most cost-effective does not mean that it is worth pursuing – it may be better than any of the others, but it may still be judged to be too costly. Cost-effectiveness is therefore a concept that is most useful in eliminating projects from consideration and in identifying ones that deserve further consideration. A project will not be implemented simply because it is the most cost-effective.

Summary

Transportation, water resource and other infrastructure-based systems serve various needs of society. One aspect of the performance of such systems is therefore how well they actually serve those needs, as measured by the cost to the user and the quality of the service that they receive.

Total, average, marginal, and incremental costs are all important aspects of system performance. The average cost is the total cost divided by the total volume of usage; the units could be the cost/vehicle for a highway or the cost/gallon for a municipal water system. The marginal cost is the cost per additional user, which would be the cost for one more car on a highway or for an additional gallon of water. Pricing and operating decisions are often based upon marginal costs rather than average costs. The incremental cost is the added cost for a larger increment of users, e.g. a 10% increase in vehicles or water usage. Investment and longer-term operating decisions are often based upon consideration of incremental costs: what is the best way to handle the expected increase in usage over the next five years? It is possible to construct engineering-based cost functions that can be used to estimate costs based upon the engineering and operating characteristics of the system.

For many systems, it is useful to consider the distinction between fixed and variable costs. Fixed costs are those associated with making the system available, while variable costs are those that vary with the level of usage. For investments in infrastructure, there are often options that can provide better service or higher capacity, but that require higher fixed costs. A common question is therefore to decide whether or not there will be sufficient demand to justify the alternative with the higher fixed costs.

Since infrastructure lasts a long time, it is important to consider costs over the entire life of the infrastructure. Design costs are likely to be modest compared to the construction costs that follow. Once the project is completed, there will costs of operation that may be borne by the owner, users or abutters. As the facility ages, it may be desirable to expand or to rehabilitate the facilities. Eventually, as a result of declining demand or excessive costs of maintenance, it will be necessary to decommission the facilities and to salvage metal or whatever is left of value. Ideally, the design of infrastructure-based systems will address total lifecycle costs, rather than simply construction costs. Adding room for expansion, allowing more efficient operations, using designs that facilitate maintenance and ensuring safer operations may require additional investment, but result in lower costs over the life of the infrastructure.

The design phase offers the greatest opportunity to affect the life cycle costs of a project. At this point, while all things will not be possible, many things will be. A major design consideration will be the extent to which the owner or developer considers the costs to users and abutters. Small changes in design conceivably produce more efficient operations or limit the negative effects on third parties, but only to the extent to which such costs are even considered in the design. Mistakes regarding the size of the project – too big, too small, too soon, too little flexibility, too difficult to rehabilitate, too much impact on neighbors – conceivably can be rectified with little or no additional time or expense related to construction.

'i.e. the profitability and the return on investment for the system. The return on investment is the annual profit divided by the total amount of the investment. While financial performance is not the only measure, or even the most important measure, if financial performance is deemed to be unacceptable then it will be difficult or impossible to find investors (or taxpayers) willing to invest more money in improving or expanding the system. The interaction between supply and demand will be an important factor determining the prices that can be charged for a service.

A third aspect of performance will be the ability of the system to handle growth in demand. The maximum capacity of a system is limited by the initial design, engineering factors, and operating constraints. The maximum capacity may be useful in design, but cannot be achieved except for brief periods. Operating capacity recognizes the importance of considering such things as the normal variations in volume, weather, and the need of routine maintenance. The operating capacity is what is achievable on most days when the system operates pretty much as planned. The limit to operating capacity is likely to be what users view as acceptable delays or restrictions on use during peak periods. The sustainable capacity is somewhat lower, because time must eventually be allowed for more maintenance and there will be periods when severe weather or accidents disrupt operations for a period of hours, days, or weeks. Sustainable capacity for most systems will be on the order of 70% of maximum capacity.

A fourth aspect of performance will be the safety and security of the system, i.e. the likelihood of accidents or disruptions to the system based upon system problems or attacks upon the system. Risk is a useful concept for considering safety and security issues, as risk is the product of two key factors: the likelihood of an accident or incident and the expected fatalities, injuries and other consequences if there is an accident or incident. Probabilistic risk assessment is a methodology that can be used to determine the cost-effectiveness of various options for reducing risk.

Figure 13 Union Pacific's Rail Freight Terminal in North Platte, Nebraska
During the late 20th century, the rail industry struggled to increase capacity to keep pace with the demand for coal and the rapid growth in the shipment of containers by rail.

Equivalence of Cash Flows

"What it comes down to is pieces of paper, numbers, internal rate of return, the net present value, discounted cash flows – that's what it's all about. ... Sure, we want to build quality and we want to build something that is going to be a statement, but if you can't do that and still have it financed and make a return, then why are we doing it?"

Terry Soderberg[1]

Introduction

Many processes are needed to assess the performance of infrastructure projects and to evaluate alternatives for improving performance. Inevitably, there will be many aspects of performance to consider and many possible impacts on society or the environment that must be minimized or mitigated. Financial analysis, which is concerned with the cash flows directly related to a project, will be critical, but so will economic analysis, which also includes the impacts of a project on the overall economy. Both financial and economic impacts can be measured in monetary terms; which types of impacts are considered will depend upon who is doing the analysis. Owners, developers and users will largely be interested in financial matters; public agencies that must approve projects are concerned with broader economic matters, such as job creation and regional prosperity.

Financial impacts concern the cash flows that are directly related to the project; these cash flows are what will determine the profitability of the project and the returns to those who invest in the project. If there is not enough cash to support construction, then a project will not be completed; if it appears that revenues will be insufficient to provide a desirable rate of return for investors, then they will not invest in the project. In Panama, the French effort to build the canal suffered from the extraordinary loss of life, but the project failed because the interest charges on the construction debt rose higher than predicted revenues – and the canal was nowhere near to being finished.

Economic impacts are broader than financial impacts, as they may include costs and benefits related to such things as consumer surplus, multiplier effects of construction, or safety, all of which are important factors to the public, even though they may not show up as revenues or expenses related to the investment in or operation of the project. Financial and economic impacts together are likely to motivate many projects, but there will also be social and environmental objectives and constraints that must be considered in proposing and evaluating projects that are aimed at enhancing the sustainability of infrastructure-based systems. The Panama Canal was eventually completed only when the U.S. government, which had strategic interests in a shorter all-water link between its east and west coasts, was able to invest what was needed to complete a less ambitious project, namely a canal with locks rather than a sea-level canal.

In this essay, the focus is on economic and financial aspects of projects, not because these are the most important measures of success, but because there are well-defined techniques that are commonly used to evaluate financial feasibility and economic desirability of a project. The concept of **net present value** (NPV) provides a very useful method for determining whether the predicted future benefits of projects justify the investment. NPV also provides a useful way for comparing different alternatives that may be proposed for a project.

When seeking ways to improve infrastructure performance, there will always be many alternatives to consider. It may be possible to improve performance by investment in markedly different types of infrastructure, by regulating land use or development or by subsidizing certain types of activities. To determine which is best, from an economic or financial perspective, it will be necessary to compare costs and benefits over a long time period. Calculating the NPV for each option provides a convenient way to make such a comparison.

---

[1] Terry Soderberg was in charge of leasing a 50-story office tower for the Worldwide Plaza, quoted by Karl Sabbagh in **Skyscraper: The Making of a Building**, Penguin Books, NY, NY, 1991, p. 377

To simplify the presentation, let's focus on financial matters.[2] For each major alternative, it is necessary to predict cash flows over a long time horizon, taking into account the costs of construction, the continuing costs of maintenance, and the costs and revenues related to operations. A typical proposal will have cash flows similar to those shown in Figure 1, which shows the net annual cash flows over the life of a hypothetical project. Net cash flows are the sum of revenues, subsidies and any other source of income minus investment, operating, maintenance and any other type of expense. In a typical proposal, cash flows are negative at the outset, because of the expenses related to planning, site acquisition, and construction. Once the project is completed, revenues start to offset continuing costs of operation and maintenance. Eventually, as the structures age or as competitors capture more of the market, net cash flows begin to decline. At the end of the life, there may be expenses related with tearing down the structure.

Figure 1 Cash Flows for a Typical Infrastructure Project

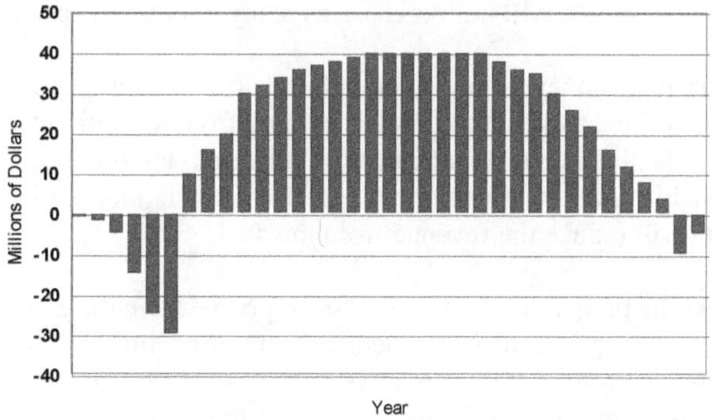

The alternatives that must be investigated may have sizable differences in terms of investments, construction costs, performance capabilities, and projected operating costs and revenue potential. Comparisons among options with markedly different cash flows will be difficult. For example, how should a low-cost option be compared with an option that requires much higher investment, but offers a chance to earn more money over a longer time frame? To answer questions like this, it is necessary to understand a basic concept of engineering economics, namely the **equivalence of cash flows**. If someone – an individual, a public agency or a company – is indifferent between two projected streams of cash flows, then those cash flows can be viewed as equivalent for that person, agency, or company. It is particularly useful to be able to take the complex flows of a typical project and compare them to something that is equivalent, but easier to understand. One obvious possibility would be to determine the amount of cash – the net present value – that would be equivalent to each projected stream of cash flows. The projected cash flows for each alternative could then be reduced to something as simple as a deposit to or a withdrawal from a bank account, either today or at some point in the future. Comparing alternatives would then be trivial, at least in terms of financial matters: the bigger the deposit the better, any deposit is better than any withdrawal, and if the best option is equivalent to a withdrawal then it is clear that there better be some non-financial objective for pursuing the project! This is why net present value is such a widely used measure of financial performance.

The next section goes into more detail concerning the time value of money, the need to discount future cash flows, the concept of a discount rate, and the concept of equivalence. Equivalence relationships can most easily be understood in the context of fixed interest payments, where there is a well-defined relationship between money invested today and the interest that will be earned over time

---

[2] Discounting and NPV analysis can be applied to any stream of costs and benefits that can be expressed in monetary terms. It is convenient to begin by focusing on the cash flows that are directly related to the project, as the cash flows are well-defined and easily understood. Moreover, for most investors, private companies, and entrepreneurs, the analysis of cash flows dominates their concerns.

Time Value of Money

Cash today is worth more than a promise that you will receive the same amount of cash in the future. There are several reasons why this is so, including the opportunities for investing the cash today, the likelihood of inflation, and the risk that the promised cash will not materialize. If the money is invested in a low-risk investment, such as savings bonds or a savings account at a bank, then the money will earn interest and the total amount available will be greater in the future. If the money is invested in stocks, bonds, or real estate, there could be even greater returns. Thus, there is an **opportunity cost** if money is only available in the future rather than being available today.

The second major reason for preferring money today rather than in the future is that **inflation** will generally reduce what can be bought with a given amount of money. Having the same amount of money in the future will not be as good as having the money today because it will not purchase as many goods and services.

The third major reason for preferring money today is **financial risk**: something could go wrong, causing the future payment to be smaller or later than expected. If the money is coming from the anticipated sale of property, there could be less than expected if there is a decline in the housing market. If the money is coming from the repayment of a loan, perhaps the borrower will be unable to make the payment. If the money is linked to some sort of international deal, perhaps a change in government will reduce the revenue from the deal.

These and other factors all affect the time value of money. Since people have different needs and expectations about the future, they will vary in their perceptions of investment opportunities, inflation, and financial risk, and different people and different organizations will put different relative values on current and future sums of money. For now, suffice it to say that there are several major reasons why money in the future is less valuable than money at hand in the present. Therefore, it is necessary to **discount** future cash flows. "Discount" comes from a Latin term that means "count for less," so discounting future cash flows means that they will count for less when evaluating a project. The **discount rate** is defined as the annual percentage by which future cash flows (a **future value**) must be reduced (discounted) to a **present value** for comparison with current cash.

The simplest way to understand discounting is to consider the benefits of investing in something safe that earns a respectable, steady $i\%$ interest per year. After one year, the money will have increased by a factor of $(1+i\%)$. If the same interest rate is maintained for t years, the money will have grown by a factor of $(1+i\%)^t$. In other words, assuming an interest rate of $i\%$, M dollars today is equivalent to $M(1+i\%)^t$ dollars in the future. To look at this same situation from the perspective of the future, M dollars in the future would be equivalent to $M/(1+i\%)^t$ dollars today if those M dollars were invested and earning $i\%$ per year.

For example, suppose you deposit \$1,000 in a bank that pays 4% interest at the end of each year. How much will you be able to withdraw at the end of five years? To determine the future value of your deposit, it is necessary to begin by calculating the annual interest that will be received at the end of the first year and adding this interest to your account. If the interest rate is 4%, then the value of the account will increase by 4% at the end of the year. The same procedure can be repeated for four more years. The results will be as shown in Table 1; the value at the end of one year equals the value at the beginning of the next year, and the value at the end of five years will be \$1,216.65. This result could also be obtained directly as $\$1000 * (1.04)^5 = \$1,216.65$.

Table 1 Future Value of Money Deposited in a Bank Account

Year	Value at Beginning of Year	Interest Rate	Interest Received at End of Year	Value at End of Year
1	$1,000	4%	$40	$1,040.00
2	$1,040	4%	$41.60	$1,081.60
3	$1,081.60	4%	$43.26	$1,124.86
4	$1,124.86	4%	$44.99	$1,169.86
5	$1,169.86	4%	$46.79	$1,216.65

It is often useful to prepare a **cash flow diagram** that depicts exactly what is being analyzed. In such a diagram, it is important to know whether cash flows occur at the beginning, middle, or end of the period. The end of one period can be assumed to equal the beginning of the next period. Figure 2 shows the cash flow diagram for Table 1. In this diagram, it is assumed that cash flows occur at the beginning of the period. From your perspective, there are just two cash transactions: a deposit of $1000 deposit is made today and a withdrawal of $1216.65 is made five years later. The annual interest payments will be added to your account and will not be taken as cash; they therefore are not shown on this chart. The chart shows the deposit at the beginning of month 1 and the withdrawal five years later at the beginning of month 61.

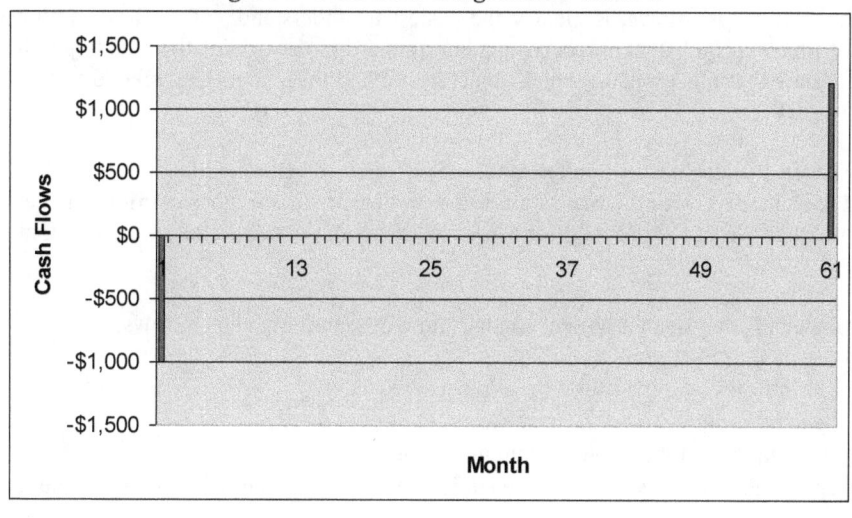

Figure 2 Cash Flow Diagram for Table 1

In financial terms, the $1,000 that is invested at the beginning of the first month is equivalent to the $1,216.65 that will be available at the end of the last month. The same chart could be used to illustrate the amount of money that must be invested today to grow to $1,216.65 in 60 months.

The choice of a discount rate is a very important issue in project evaluation, because large investments in the near future must be justified by benefits that are achieved over a long time period. The more that those benefits are discounted, the harder it is to justify the investment. Table 2 illustrates the present value of $100 received five to 100 years in the future and discounted at a rate of 1 to 20%. The higher the discount rate or the longer the period, the lower the present value.

Table 2 Present Value of $100 Received in 5, 10, 20 or 50 Years

Discount Rate	5 Years	10 Years	20 Years	50 Years
1%	$95	$91	$82	$61
5%	78	61	38	8.80
10%	62	38	15	0.90
20%	40	16	2.60	0.11

Equivalence Relationships

Equivalence is a key concept in project evaluation and project finance. Two projected streams of cash flows are equivalent for someone if they are equally acceptable, i.e. if the individual is indifferent to receiving one or the other. As noted in the introduction, our goal is to transform an arbitrary stream of cash flows into an equivalent cash flow that is easily understood, such as present value, future value, or uniform annual value.[3] Given a discount rate, it is possible to calculate these three measures so that each is equivalent to a given stream of cash flows:

1. Present value or present worth (P): the equivalent present value of the cash flows (what is the projected stream of cash flows worth today?)
2. Future value or future worth (F): the equivalent future value of the cash flows (what is it worth at a specified time in the future?)
3. Annual value or annuity worth (A): the equivalent annuity amount (what is it worth in terms of receiving a uniform cash flow of A at the end of each period for a specified number of periods).

P, the present value of the cash flows, is clearly the easiest to understand. If we have a choice among various alternatives, each of which is equivalent to receiving a lump sum of money today, then the larger the present value the better (assuming for now that money is our chief object in life and that the money related to the various projects is indeed legally acquired!).

The other two measures are also easy to understand. If our time of reference is some time in the future, we will presumably want the alternative with the maximum future value. If we are more comfortable dealing with monthly or annual cash flows, then we can express options in terms of an annuity and choose the one that pays the most per period.

The basic tasks in financial analysis of a project can therefore be summarized as follows:

- Predict the cash flows over the life of the project
- Estimate the net present value of the cash flows
- Calculate the equivalent future value or annuity, if desired
- Rank projects by P, F, or A (since they are equivalent measures, the ranking will be the same)

Let's assume that we are a private company considering various projects that we can finance by borrowing from a bank at an interest rate of i%. For simplicity, assume that we can also put money into a savings account at the bank and also receive i% interest. Our choices therefore are either to invest in more projects or to put money into our bank account. If a project earns more than i%, then we will borrow money from the bank; if it earns less than i%, we will be better off putting money into our bank account.

It is useful to introduce some notation for discounting. [F/P,i,N] can be used to denote the factor that calculates the future value F as a fraction of the present value P, assuming a discount rate of i% for N periods. [P/F,i,N] denotes the factor that is used to determine the present value P given the future value F, again assuming a discount rate of i% for N periods. The two factors are used as follows to compare future and present values:

(Eq. 1) $\qquad F = P * [F/P,i,N] = P * (1+i)^N$

(Eq. 2) $\qquad P = F * [P/F,i,N] = F / (1+i)^N$

It is a small step from discounting one future payment to discounting all anticipated cash flows over a project's time horizon. If you have a spreadsheet, you can use these equations repeatedly to convert an arbitrary cash flow into either a present or a future value. The present value of the entire stream of cash flows will be the sum of the individual discounted cash flows.

---

[3] Present value, future value, and equivalent uniform annual value are the terms most commonly used in general business. Present worth PW, future worth FW, and annuity worth AW are encountered in many engineering economics textbooks. This text uses the two sets interchangeably; the very common use of net present value may make the use of present and future value more desirable.

The term **net present value** is commonly used to denote the present value of a stream of cash flows that includes both costs C(t) and benefits B(t) over a designated time horizon. Given a discount rate i, it is straightforward to calculate the net benefits during any period, the present value of those benefits, and the net present value (NPV) of the entire project:

(Eq. 3)   Net benefits during period t = B(t) – C(t)

(Eq. 4)   Present value of net benefits in period t = NPV(t) = (B(t) – C(t))/(1 + i)^t

(Eq. 5)   NPV(project) = Σ((B(t) – C(t))/(1 + i)^t) for the life of the project

Sometimes it is desirable to consider an annuity rather than a present or future value. An annuity can be compared to other measures reported annually, such as revenue, operating costs, or profitability. There are multiple ways to find the equivalent annuity. Since an annuity of A per period is certainly one possible cash flow, you can find the annuity that is equivalent to either a present or future value using the above equations. If you make interest rate, annuity amount, and the number of periods a variable, you can easily find the annuity amount that is equivalent to any present or future value. However, it can be more elegant (and less time-consuming) to use algebraic expressions to convert P or F into annuities (or to convert annuities into P or F).

The equivalent future value F of an annuity A is the amount that would be accumulated by the end of the last payment assuming that all payments were invested so as to grow at a rate equal to the discount rate i. As was the case in calculating the relationships between P and F, there will be an equivalence factor that can be used to find F as a function of A. This factor depends upon the discount rate per period and the number of periods. It is denoted [F/A,i,N], and it is called the **uniform series compound amount factor**. It is assumed that a payment of A is made at the **end** of each period and that each payment is invested at i% per period for the remaining periods.

Given i and N, the future value will be proportional to A, and [F/A,i,N] will be the proportionality factor:

(Eq. 6)   F = A * [F/A,i,N]

There is a simple algebraic expression for [F/A,i,N]:

(Eq. 7)   [F/A,i,N] = [(1+i)^N-1]/i

Values for the expression can be found in tables, and spreadsheets have functions that will calculate the expression for you.

<u>Derivation of the Uniform Series Compound Amount Factor [F/A,i,N]</u>

The derivation of this expression for [F/A,i,N] is based upon well-known relationships for geometric sequences. First, we can calculate F just by converting each payment A(t) to a value at the end of the Nth period. The payment at the end of period t will earn interest for another N-t periods, so its future value will be:

(Eq. 8)   A(t) = A(1+i)$^{N-t}$

Summing over the entire N periods:

(Eq. 9)   F(N) = A ((1+i)$^{N-1}$ + (1+i)$^{N-2}$ + ... +(1+i)$^{N-t}$ + ... + (1+i)^0)

If we let b = 1+i, this is equivalent to a simple geometric sequence:

(Eq. 10) $F(N) = A (b^{N-1} + b^{N-2} + ... + b^{N-t} + ... + b^0)$

If we rearrange the terms, then

(Eq. 11) $F(N) = A (1 + b + ... + b^{N-t} + ... + b^{N-1}) = A [F/A,i,N]$

And

(Eq. 12) $[F/A,i,N] = (1 + b + ... + b^{N-t} + ... + b^{N-1})$

Now use a mathematical trick: multiply this by one expressed as $(1-b)/(1-b)$ to get a more elegant result:

(Eq. 13) $[F/A,i,N] = [1/(1-b)] [(1 + b + ... + b^{N-1}) - (b + b^2 + ... + b^{N-t+1} + ... + b^N)]$

(Eq. 14) $[F/A,I,N] = [1/(1-b)] [1-b^N] = [1-b^N] / (1-b)$

Substitute $(1+i) = b$ and rearrange terms to get the uniform series compound amount factor:

(Eq. 15) $[F/A,i,N] = [(1+i)^N - 1]/i$

Expressions for the Other Equivalence Factors

Expressions for the other factors can readily be found. Since $F(N) = A * [F/A,i,N]$ we can invert the above to get $[A/F,i,N]$, which is known as the **sinking fund factor**:

(Eq. 16) $[A/F,i,N] = i/[(1+i)^N - 1]$

A sinking fund may be established by a local government or a company as a means of paying off a future debt. An amount is paid each year into the sinking fund, which could be a savings account or another safe investment. Each year, the sinking fund would earn interest, and at the end of the time period, enough would have accumulated to pay off the debt (or fix the roof on the town hall or deal with whatever problem the sinking fund was established to solve). The relevant question is the size of the annuity.

A related question involves the present worth of an annuity. This amount of money that will be available in the future if a specified amount is invested each period for n periods at interest rate i. This amount can be calculated using what is known as the **uniform series present worth factor** and symbolized as $[P/A,i,N]$. This factor represents the present value of the annuity, and the annuity has an equivalent future value represented by $[F/A,i,N]$ (Equation 15). Taking the present value of that future value will produce the desired uniform series present worth factor:

(Eq. 17) $[P/A,i,N] = [F/A,i,N] / (1+i)^N$

The uniform series present worth factor can be obtained by substituting for $[F/A,i,N]$ using equation 15:

(Eq. 18) $[P/A,i,N] = [(1+i)^N - 1]/[i (1+i)^N]$

The inverse of this expression will give the **capital recovery factor**, which can be used to determine the size of an annuity that is required to recover an initial capital investment:

(Eq. 19) $[A/P,i,N] = [i (1+i)^N]/[(1+i)^N - 1]$

These expressions may not seem too elegant, nor are they easy to remember. However, note that when N gets large, they become very clear and simple:

(Eq. 20) $[P/A,i,N] = 1/i$

(Eq. 21) $[A/P,i,N] = i$

These approximate expressions for the sinking fund and the capital recovery factors can be very useful in estimating the present value of a long-term annuity or in estimating the present value of a long-term annuity. For example, suppose a toll road generates $1 million per year in profit. What would that be worth to a potential purchaser with a discount rate of 10%? We could approach this by assuming a life of 30, 40 or 100 years and looking up the values for [P/A,10%,30], [P/A,10%,40] and [P/A,10%,100]. If we did this, we would find the following:

($1 million) [P/A,10%,30] = $1 million (9.4269) = $9.4 million

($1 million) [P/A,10%,40] = $1 million (9.7791) = $9.8 million

($1 million) [P/A,10%,100] = $1 million (9.9993) = $10 million

If we had just used the approximation, we would immediately have said that

($1 million) (1/0.1) = $10 million

This result is very close for 40 years and almost exact for 100 years. In many analyses, particularly preliminary analyses where few if any of the numbers are precise, the approximation will be quite adequate.

In going from present value to annuities, we find a similar result. In this case, the question would concern the annual profit that would be required to justify an investment of $10 million in a turnpike. The approximation says that the long-term annuity would be approximately $1 million multiplied by the discount rate of 10% or $1 million per year. The more precise calculations would call for somewhat higher returns, but nothing markedly greater than the quick estimate of $1 million:

$10 million * [A/P,10%,30] = $10 million (.1061) = $1.06 million

$10 million * [A/P,10%,40] = $10 million (.1023) = $1.02 million

Using these approximations is sometimes called the **capital worth method**.

Figure 3 summarizes the concept of equivalence. The chart in the upper left shows a typical stream of cash flows for a project. There are substantial investments during the first four years, profitable operations beginning in year five, a dip in earnings midway through the life of the project reflecting the need to expand or rehabilitate the project, and ultimately a decline in profitability and a decommissioning expense. It is impossible to tell for sure how good this project simply by looking these cash flows. The other three charts show the equivalent present value, annuity, and future value, each of which is easy to understand. Since the present value is positive, then investing in this project is better than investing in something that earns interest equal to the discount rate. If the present value is negative, then the project is not as good as investing in something that earns interest equal to the discount rate.

Figure 3 For a given discount rate, any arbitrary stream of cash flows will be equivalent to a present value, a future value or an annuity

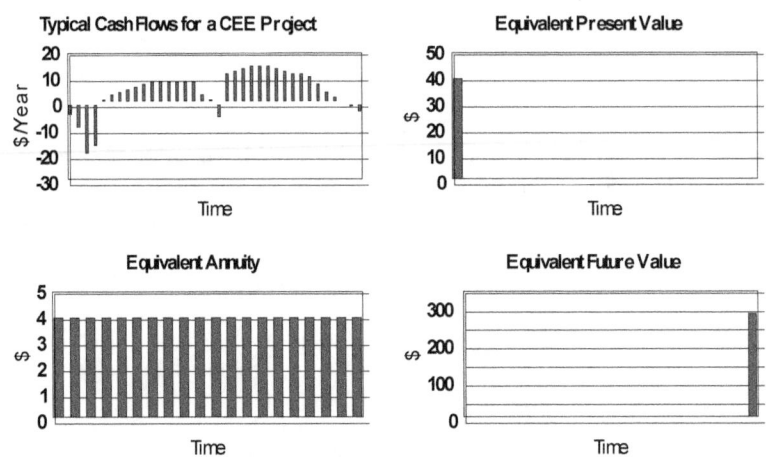

Table 3 summarizes the six factors derived above. F refers to the future value, P to the present value, and A to the equivalent annuity amount. The equations are all functions of the discount rate i and the number of periods N. The final two rows of this table highlight the two easily remembered factors known as the "Capital Worth Method." When N is large, these factors provide an easy way to get a quick estimate of the value of an annuity (A/i) or the annuity that is equivalent to any present amount (Pi).

Table 3 Summary of Equivalence Factors, Discrete Compounding

Symbol	Name	Comment	Value
[F/P,I,N]	Future value given present value	How much growth can be expected	$(1+i)^N$
[P/F,I,N]	Present value given future value	Discounted value of a future amount	$1/(1+i)^N$
[F/A,I,N]	Uniform series compound amount factor	If I save some each period, how much will I accumulate?	$[(1+i)^N-1]/i$
[A/F,I,N]	Sinking fund payment	How much must I save each period to meet my retirement goals?	$i/[(1+i)^N-1]$
[A/P,I,N]	Capital recovery factor	What will my mortgage payment be?	$[i(1+i)^N]/[(1+i)^N-1]$
[P/A,I,N]	Uniform series present worth factor	If I can pay A per month, how large a mortgage can I afford?	$[(1+i)^N-1]/[i(1+i)^N]$
[A/P,I,infinity]	Capital recovery factor for very long time periods	Capital Worth Method	i
[P/A,I,infinity]	Uniform series present worth factor for very long time periods	Capital Worth Method	$1/i$

The next two figures illustrate the relationship between annuities and present or future worth. Figure 4 shows the "uniform series, compound amount factor", which is the amount by which the future value exceeds the annual amount of an annuity depending upon the length of time and the earnings rate. This is the factor that is useful in determining how much should be invested each year toward retirement. Figure 5 shows the annuity that is equivalent to a present value for various time periods and interest rates. This factor is called the "uniform series, capital recovery factor", and an example would be the amount that is paid each year on a mortgage.

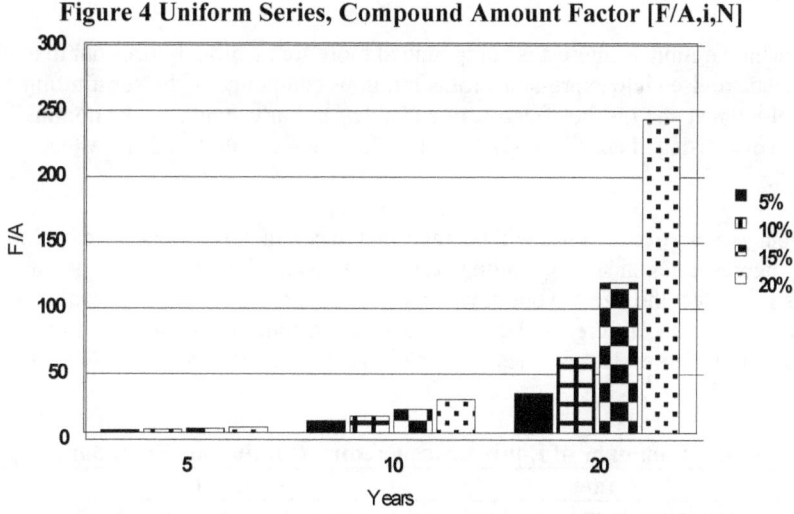

Figure 4 Uniform Series, Compound Amount Factor [F/A,i,N]

Figure 5 Uniform Series, Capital Recovery Factor [A/P,i,N]

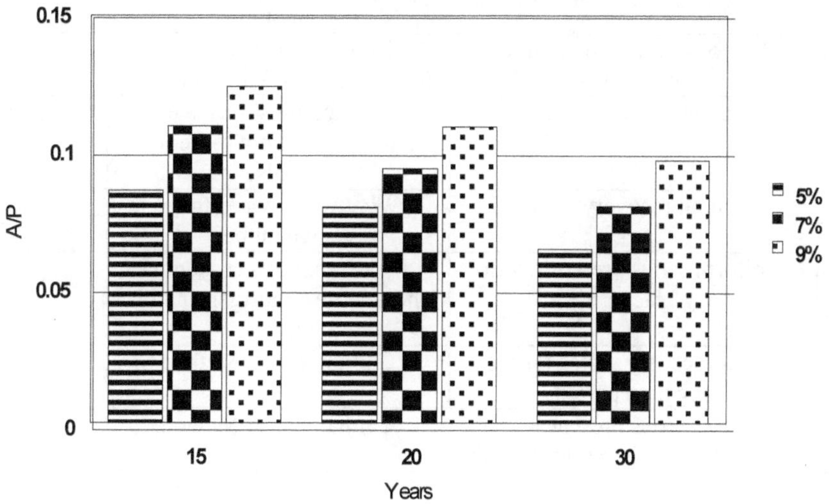

Continuous Compounding: Nominal vs. Effective Interest Rates

Interest rates are normally expressed in terms of annual returns, and the **nominal rate of interest** is the interest rate that you would receive if interest were compounded annually. If interest is compounded more frequently, then there will be a higher **effective rate of interest.** With a 12% nominal rate, an investment of $1000 on January 1 would earn $120 interest on December 31. However, if you compound interest semiannually, then you would get more interest. First after 6 months, you would earn 6% or $60 on your initial investment. Then, at the end of the year, you would earn another $60 on the initial investment plus $3.60 on the interest that you received at the end of June. For the year, you therefore earned an effective rate of 12.36% interest on your investment, even though the nominal rate was 12%.

More rapid compounding gives a bit higher effective rate, but there are diminishing returns:

- Quarterly rate is 12.55%

- Bimonthly rate is 12.62%
- Monthly rate is 12.68%
- Daily rate is 12.75%

Clearly, this is approaching a limit as interest is compounded more frequently. It turns out that all of the formulas for discrete cash flows can be revised into expressions for continuous compounding by substituting e^{rN} for $(1+i)^N$, where r is the nominal rate of interest and i is the effective rate (Table 4). For example, if the nominal rate is 12%, as in the example we have been discussing, then $e^{rN} \sim 2.718^{0.12} = 1.1275$ which equals 1 plus the effective daily rate of 12.75% that we just calculated.

The nuances of nominal versus effective rates will be important in certain areas, notably in banking and finance, where contracts will specify interest rates and compounding periods. If you are buying a car or making a deposit to a savings account, these details will affect the size of your payments or your earnings. In project evaluation – especially during the early stages of the process – the differences between discrete and continuous compounding will be minor compared to the uncertainties in estimating costs, revenues, time periods, and other factors that will affect the outcome of the analysis.

Table 4 Summary of Equivalence Factors, Continuous Compounding

Symbol	Name	Comment	Value
[F/P,I,N]	Future value given present value	How much growth can be expected	e^{rN}
[P/F,I,N]	Present value given future value	Discounted value of a future amount	$1/e^{rN}$
[F/A,I,N]	Uniform series compound amount factor	If I save some each period, how much will I accumulate?	$[e^{rN}-1]/r$
[A/F,I,N]	Sinking fund payment	How much must I save each period to meet my retirement goals?	$r/[e^{rN}-1]$
[A/P,I,N]	Capital recovery factor	What will my mortgage payment be?	$[r(e^{rN})]/[e^{rN}-1]$
[P/A,I,N]	Uniform series present worth factor	If I can pay A per month, how large a mortgage can I afford?	$[e^{rN}-1]/[r(e^{rN})]$
[A/P,I,infinity]	Capital recovery factor for very long time periods	Capital Worth Method	R
[P/A,I,infinity]	Uniform series present worth factor for very long time periods	Capital Worth Method	1/r

Some Useful Approximations

The continuous compounding formulations for [F/P,r,N] and [P/F,r,N] are very useful because they are easy to remember and can readily be used for quick approximations. Since $[F/P,r,N] = e^{rN}$ and $[P/F,r,N] = 1/e^{rN}$, it can be very helpful to remember a few useful results:

- If rN = 1, e^{rN} = 2.718...
- If rN = 0.7, e^{rN} = 2.013..., approximately 2
- If rN = 1.1, e^{rN} = 3.004..., approximately 3
- If rN = 1.4, e^{rN} = 4.055..., approximately 4

You can use these relationships to figure out how long it will take to double, triple or quadruple your money. If rN = 2, then money invested at r% for N years will double in value. If the nominal interest rate is 10%/year, then it will take seven years to double your money; if the nominal rate is 5%, then it will take 14 years to double your money. Likewise, if rN = 1.1, then $e^{rN} \sim 3$ and the future value will be about three times the present value. Thus, if you earn 10% per year for eleven years, you can triple your money. You can also use the inverse relationship to calculate present values. For example, if the nominal interest rate is 7%, then the present value of something received in ten years will be half its future value.

In short, it is possible to make mental estimates of quite complex functions made up of incomprehensible expressions such as $(1+i)^n$. Mental math has been an increasingly undervalued skill since the invention of the electronic calculator, but you will certainly find it useful to use the above relationships if you are ever involved in face-to-face discussions, negotiations, or debates related to project evaluation when it would be inconvenient or inappropriate to use your calculator or computer. You <u>can</u> do present value analysis in your head - and that <u>will</u> give you an advantage in negotiation!

<u>Proof that $e^{rN} = (1 + i)^N$</u>

To prove this relationship, we can begin by writing an algebraic expression for what we are doing as we compound more frequently. If r is the nominal interest rate, but we compound M times per year, then the effective rate will be

(Eq. 22) $\qquad i = [1 + (r/M)]^M - 1$

and the factor [F/P,<u>r</u>%,1] will be

(Eq. 23) $\qquad [F/P,\underline{r}\%,1] = [1 + (r/M)]^M$

Note that the term "<u>r</u>%" represents the nominal interest rate r and the use of continuous compounding.

If we let p = M/r and rewrite this equation, we get

(Eq. 24) $\qquad [F/P,\underline{r}\%,1] = (1 + 1/p)^{rp} = ((1 + 1/p)^p)^r$

This is a classic relationship, as the limit of $(1+1/p)^p$ as p approaches infinity is e = 2.7128...!
Thus, we have

(Eq. 25) $\qquad [F/p,\underline{r}\%,1] = e^r$

and therefore

(Eq. 26) $\qquad [F/P,\underline{r}\%,N] = e^{rN}$

which somewhat unexpectedly gives us a very nice relationship:

(Eq. 27) $\qquad [F/P,\underline{r}\%,N] = e^{rN} = (1 + i)^N$

where the exponential expression assumes continuous compounding using the nominal rate and the other expression uses the effective rate. Hence, we can revise all of the formulas for discrete cash flows into expressions for continuous compounding by substituting $e^{rN} = (1 + i)^N$.

Financing Mechanisms

Equivalence relationships enable financing of large projects. Those investors or banks that have money to invest are willing to make cash available for implementing a project in return for future interest payments, mortgage payments, or dividends. This section describes financing mechanisms that are used for infrastructure projects.

Mortgages

A mortgage is a loan that is backed by property. If the borrower defaults on a payment, then the lender can seize the property and either use it or sell it to recover their costs. From the lender's perspective, a mortgage is less risky than a long-term unsecured loan, and therefore merits a lower interest rate. A mortgage will typically be limited to about 80% of the assessed value of the property in order to reduce the risk to the lender in case the owner defaults. If the mortgage is less than the value of the property, then the bank will be able to sell the property and regain its investment.

The monthly payments on a mortgage depend on the amount of the loan (the principal amount), the interest rate and the period of the mortgage. If the mortgage principal is PRICE, the annual interest rate is i% and the term is N years, then the monthly payment will be:

(Eq. 28)   M = (PRICE) [A/P,i%,N]

where $[A/P, i\%, N] = [i (1+i)^N]/[(1+i)^N - 1]$ as shown in Table 3. This factor can be found in several ways. It can of course be calculated using the formula, but it can also be obtained from a table (e.g. Table 5) and it can be obtained using a function available on many spreadsheets. In Excel, the function PMT(i%,N,P) will give the desired answer.[4]

Table 5 Capital Recovery Factor [A/P,i%,N] for selected interest rates i% and years N

Years	3%	4%	5%	6%	7%	8%	9%	10%
5	0.2184	0.2246	0.2310	0.2374	0.2439	0.2505	0.2571	0.2638
10	0.1172	0.1233	0.1295	0.1359	0.1424	0.1490	0.1558	0.1627
15	0.0838	0.0899	0.0963	0.1030	0.1098	0.1168	0.1241	0.1315
20	0.0672	0.0736	0.0802	0.0872	0.0944	0.1019	0.1095	0.1175
25	0.0574	0.0640	0.0710	0.0782	0.0858	0.0937	0.1018	0.1102
30	0.0510	0.0578	0.0651	0.0726	0.0806	**0.0888**	0.0973	0.1061

For example, let's calculate the annual and monthly payments on a 30-year mortgage for $1,000,000 at 8% interest. The mortgage payment will be the annuity that – for the bank – is equivalent to the principal amount of the loan. To obtain the annual payment, the amount of the loan needs to be multiplied by the capital recovery factor [A/P,i%,N], which Table 5 shows to be 0.0888 for a 30-year mortgage with an 8% interest rate,. The annual payment would therefore be approximately $88,800:

(Eq. 29)   PMT = (PRICE) [A/P, 8%, 30] = $1,000,000 (0.0888) = $88,800

If payments were to be made monthly, then the PMT would be approximately 1/12th of the annual amount or $7,333.33. These amounts are approximate because the table only shows the capital recovery factor to four significant digits. More accurate answers for annual or monthly mortgage amounts could be obtained by using the formula for the capital recovery factor or more readily by using the PMT function in Excel:

(Eq. 30)   Annual payment = PMT(8%, 30, $1,000,000) = $88,827.43

---

[4] The Excel function is PMT(interest rate, number of periods, present value). Tables such as Table 5 can easily be created in any spreadsheet.

To obtain the monthly payment, it is necessary to use the monthly interest rate and 360 monthly payments. Using the PMT function, the result is:

(Eq. 31)   Monthly payment = PMT(8%/12, 360, $1,000,000] = $7337.65

These results are slightly higher than the results obtained using the factor from Table 5. The bank would be sure to use the actual number, but the difference is far too small to make any difference in project evaluation, which naturally must deal with many ill-defined numbers when comparing options. Neither the amount of the loan nor the interest rate would be known with certainty until a particular option has been chosen and the project is almost completed.

After making mortgage payments for several years, an owner may decide to refinance the mortgage. The most likely reasons for refinancing would be to reduce monthly payments or to obtain cash:

- Interest rates may be lower, so that the mortgage payment could be reduced.
- The mortgage could be extended over a longer time period in order to reduce monthly payments.
- It may be possible increase the size of the mortgage and extend the term of the mortgage without increasing the size of the monthly payment.
- If the value of the property has increased, the owner may be able to borrow additional money for another project.

There are three steps in refinancing:

1. Figure out the remaining balance on the initial loan.
2. Negotiate terms for the new loan.
3. Close on the new mortgage:
   a. Use some of the proceeds to pay off the original loan
   b. Give a check to the mortgagee for any additional funds that are borrowed.

The remaining balance on the new loan can be calculated as of a point in time immediately after a regular mortgage payment has been made. The remaining balance could be calculated in multiple ways:

1. Read the statement: the monthly or annual statements will show the amount outstanding as of the previous payment, the amount due for the current payment, and the portion of the payment that goes toward interest and principal. This is what homeowners look at when they are considering refinancing.
2. Calculate the value of all the payments that have been made so far and subtract from the original amount of the loan.
3. Calculate the value of the remaining payments, using the interest rate of the loan as a discount rate. This is equivalent to what must be repaid to the bank.

These calculations will involve multiple applications of various equivalence factors. For example, suppose that payments had been made for 15 years on the 30-year, 8% mortgage for $1 million that was described above. If the borrower has a chance to refinance with a 20-year mortgage carrying a 6% interest rate, what would the new payments be? First, it is necessary to calculate the amount remaining on the mortgage. Assume that the borrower just made the 15th payment, so that there would be 15 more payments due on the original 30-year mortgage. Although half the payments have been made, much less than half the loan has been paid off, because most of the payments have gone toward interest.

There are various ways to calculate the amount remaining. One way would be to subtract the amounts paid off each year from the initial purchase price. The first payment of $88,827 included $80,000 interest (8% of $1 million) and therefore a principal payment of $8,827. With a spreadsheet, it is possible to continue the analysis, reducing the remaining principal by the proper amount after each payment as shown in Table 6. The amount remaining after the 15th year is the amount shown at the beginning of year 16, namely $760,317.

Table 6 Payments of Principal and Interest and Remaining Balance Over the Life of a 30-year Mortgage for $1 million at 8% Interest

Year	Mortgage balance	Payment	Interest	Principal
0	$1,000,000	$88,827	$0	$0
1	$1,000,000	$88,827	$80,000	$8,827
2	$991,173	$88,827	$79,294	$9,534
3	$981,639	$88,827	$78,531	$10,296
4	$971,343	$88,827	$77,707	$11,120
5	$960,223	$88,827	$76,818	$12,010
6	$948,213	$88,827	$75,857	$12,970
7	$935,243	$88,827	$74,819	$14,008
8	$921,235	$88,827	$73,699	$15,129
9	$906,106	$88,827	$72,488	$16,339
10	$889,767	$88,827	$71,181	$17,646
11	$872,121	$88,827	$69,770	$19,058
12	$853,063	$88,827	$68,245	$20,582
13	$832,481	$88,827	$66,598	$22,229
14	$810,252	$88,827	$64,820	$24,007
15	$786,244	$88,827	$62,900	$25,928
16	$760,317	$88,827	$60,825	$28,002
17	$732,314	$88,827	$58,585	$30,242
18	$702,072	$88,827	$56,166	$32,662
19	$669,410	$88,827	$53,553	$35,275
20	$634,136	$88,827	$50,731	$38,097
21	$596,039	$88,827	$47,683	$41,144
22	$554,895	$88,827	$44,392	$44,436
23	$510,459	$88,827	$40,837	$47,991
24	$462,468	$88,827	$36,997	$51,830
25	$410,639	$88,827	$32,851	$55,976
26	$354,662	$88,827	$28,373	$60,454
27	$294,208	$88,827	$23,537	$65,291
28	$228,917	$88,827	$18,313	$70,514
29	$158,403	$88,827	$12,672	$76,155
30	$82,248	$88,827	$6,580	$82,248
31	$0			

A more elegant approach is to think about the value of the remaining payments rather than worrying about the contributions so far to principal and interest. From the lender's perspective, they are receiving 15 more payments at the original 8% interest rate. The value of this annuity, discounted at 8%, will also give the amount remaining on the mortgage:

(Eq. 32)   Outstanding amount = $88,000 * [P/A, i%,15] = $88,800 * 8.5595 = $760,084

This amount is slightly less than the balance of $760,317 shown in the table due to rounding errors. With either answer, we can express the outstanding amount as $761 thousand and estimate the new mortgage payment as just over $66 thousand:

(Eq. 33)   New payment = $761,000 * [A/P, 6%, 20] = 761,000 * .0872 = $66,359

Thus, by refinancing, the borrower can reduce the annual mortgage payment by $22 thousand. The annual payments are lower because of the lower interest rate and also because the repayment period has been extended by 5 years (15 remaining years on the original mortgage vs. 20 years on the new mortgage).

This example illustrates how multiple paths may be used to reach the same answer. The equivalence relationships can be used repeatedly, in different ways, and so long as the logic is correct along each path, they will each reach the correct destination.

Bonds

Bonds provide a way for companies or agencies to raise money. A bond is offered with a face value V, a life of N years, and annual interest of i%. Interest payments are made at the end of each year for N-1 years. At the end of the Nth year, the owner receives the final interest payment and the bond is redeemed for its original face value. Bonds are commonly sold in denominations of $1000 with a 30-year term, but other options are available. Bonds can be bought and sold over their lifetime, so it is possible to buy a 30-year bond that will become due in less than 30 years. Bonds can even be offered as **zero coupon bonds**, in which the seller pays no interest but promises to pay the face value at the end of the term; for these bonds, the purchase price will be much less than the face value.

Bonds are supported by the credit of the issuing agency or company; if the company fails to pay interest when it is due, then the bond-holders can force the company into bankruptcy. If the company declares bankruptcy, then the assets of the company are divided up among the creditors, including the bondholders. Various credit agencies rate the quality of bonds, so that investors have a reasonable idea of the risks involved in buying the bonds. The higher the perceived risks of the issuing agency or company, the lower the credit rating on the bonds, and the higher the interest rates that must be offered to attract investors.

Three interesting questions are 1) the value of the bond to an investor and 2) the change in the value of the bond as interest rates change, and 3) the change in the value of the bond as perceived risks associated with the issuing agency or company change. The value of the bond to an investor depends upon the investor's perception of the risks involved, the investor's discount rate for bonds with such risks, and the interest rates offered for the bond. If the investor's discount rate is lower than the interest rate that is offered, the investor will consider buying the bond. Market forces related to the supply and demand for fixed interest securities will determine what interest rates are actually required to sell bonds.

Assume that the bond is sold at time 0. The seller agrees to pay interest of i% per year for N years and, at the end of N years, to pay back the face value of the bond. The seller offers the bonds to the marketplace, and potential purchasers decide whether or not they want to buy the bonds. The purchasers may plan to hold the bonds until maturity, or they may merely view the bonds as a short- or medium-term investment. The value of a high quality bond to an investor can be calculated as follows:

(Eq. 34)   Value = Annual Interest [P/A, i%, N] + Face Value [P/F, i%, N]

The first term is the present worth of the N annual interest payments and the second term is the present worth of the final redemption of the bond at its original face value. It is crucial to recognize that the investor's discount rate of i% can be higher or lower than the interest rate on the bond. If there is a possibility that the bond will default, both terms could be reduced by a factor representing the probability that interest or the final redemption would not be made.

Assuming that the probability of default is close to zero, the value of a bond to an investor will depend upon the face value, the years to maturity, the interest rates, and the discount rate of the investor. If the face value is $1,000 and the interest rate is 6% for a 30-year bond, then the value to a potential purchaser who also has a 6% discount rate will be exactly $1,000:

(Eq. 35) Value = ($1,000) (0.06) [P/A, 6%, 30] + $1,000 [P/F, 6%, 30]
= $60 (13.7648) + $1,000 (0.1741)
= $826 + $174 = $1,000

For someone with a discount rate of 5%, the bond will be worth much more:

(Eq. 36) Value = $60 (15.3725) + $1,000 (0.2314)

= $922 + $231 = $1,153

Likewise, someone with a discount rate of greater than 6% would value the bond at less than $1,000. In each case the equivalence factors were obtained from a table of equivalence factors.

The same type of calculations can be used to show how the value of a bond could change if interest rates change. For example, suppose that 20 years have gone by and interest rates on similar bonds have fallen to 5%. The 6% bond is therefore paying more interest than a bond with similar risk would have to pay today; hence, investors would find that bond more appealing. Someone with a 5% discount rate would be willing to pay $1,000 for the bond paying 5%, but would pay more for the 6% bond with 10 years until maturity:

(Eq. 37) Value = $60 [P/A, 5%, 10] + $1000 [P/F, 5%, 10]

= $60 (7.7217) + $1,000 (0.6139)

= $463 + $614 = $1077

Thus, when interest rates fall, bond values rise. In this instance, note that the redemption value is now much greater than the value of the interest payments. As the bond approaches maturity, the redemption value dominates.

Sinking Funds

A sinking fund can be established to meet expected future capital needs, such as paying the principal on bonds when they become due or conducting a major rehabilitation of a factory at some distant point in the future. Companies or agencies can pay a constant amount into a fund that is maintained solely to cover this future capital need. As shown above, the sinking fund factor [A/F,i%,N] can be used to determine the annual amount A to invest at i% to reach the goal of having an amount F at the end of N periods.

The calculations will require another step if the future need is itself an annuity. For example, planning for retirement involves consideration of three questions. First, how much will be needed in annual income after retirement. Second, how much will be needed in savings to produce this level of retirement income. Third, how much will need to be saved each year (e.g. in a sinking fund) in order to have accumulated the money that will be used to purchase the retirement annuity. The first question is a matter of personal needs and desires; in effect, it is necessary to determine how large an annuity will be needed to support your desired lifestyle in retirement. The second question is a matter of equivalence: how much will be needed to purchase the retirement annuity? This is the retirement goal. The third question is a different matter of equivalence: how much do you have to save each year to reach your retirement goal? The relevant equations are as follows:

(Eq. 38) Retirement Goal = Retirement Annuity [P/A,i%,N]

(Eq. 39) Savings Goal = Retirement Goal [A/F,g%, M]

The key unknowns are a) how much will really be needed, b) what interest rate (i%) can you expect for the annuity, and what annual return on investment (g%) can you expect on your savings. Perhaps the most interesting variable for many people will be M, the years until retirement. The more that is saved, the higher the return on investment, and the higher the interest rate on the retirement annuity, the sooner you can retire. The longer you work the more time you will have to reach your retirement goal, and the less you will need to save each year to reach that goal.

Consider a 30-year old couple planning for retirement. They would like to save a constant amount per year in a tax-deferred investment account. After working for another 30 years, they hope to have accumulated enough funds to provide an annuity that will last them until they are 100 years old or older. If they can achieve earnings of 8% per year on their investments, how much should they save each year in order to be able to have $80,000 per year if they retire at age 60 and live forever?

To answer this, assume a discount rate of 8% for all of the calculations. While it is unlikely that they will live forever, that hope at least makes the analysis a bit easier: the anticipated 8% return on their accumulated retirement investments will be $80,000, so they will need to accumulate $80,000/0.08 = $1 million by age 60 (note that this approach uses the simple capital worth method to convert the desired annuity into a required sum). The question therefore is what level of annual investment is needed to accumulate a million dollars in 30 years if returns are 8%?

(Eq. 40) Annual investment = $1,000,000 [A/F, 8%, 30] = $1,000,000 * (.0088) = $8,800

This may seem steep. If they extend their planned retirement ten years to age 70, then they will only need to invest $3,900 per year:

(Eq. 41) Annual investment = $1,000,000 * [A/F, 8%, 40] = $1,000,000 * .0039 = $3,900

Toll-Based Financing

For highways and bridges that carry a lot of traffic, it may be possible to use the projected toll revenues to justify issuing bonds that are sufficient to cover the cost of construction. Some quick estimates may indicate whether or not toll-based financing will work. The interest on the bonds must be compared to the net revenue from the tolls, which is what is left after paying for the annual operating costs of the bridge. If the net toll revenue is well above the anticipated interest payments, then this would be a good candidate for toll-based financing.

Consider a proposal for a bridge that is expected to cost $50 million to construct and $3 million per year for operating and maintenance costs. The bridge is expected to serve five to ten million vehicles per year, and the governor believes that the public would view a toll of $1 to $2 as reasonable. Will the tolls cover the cost of bonds that could be sold to pay the construction costs of the bridge? Interest rates on the bonds are expected to be 4%.

Given the projected traffic volume, the bridge would earn $5 to $20 million per year with a toll of $1 or $2. After deducting $3 million per year for operating expenses, the net revenue would $2 to $17 million. Give the interest rate of 4%, the annual interest on the $50 million investment would be $2 million. Thus, even the lowest estimate of net revenue would cover the annual interest payments on bonds. With a $2 toll, funds would be available for related projects, such as improving access roads or providing support to other transportation projects.

Summary

When evaluating projects, it will be necessary to compare costs and benefits that are incurred over a period of many years. The costs of the project are generally concentrated at the outset of the project, as benefits do not begin until construction is at least partially completed. One basic question for any proposed project is whether or not the eventual benefits will be sufficient to justify an initial investment in a project. In most situations, there will be multiple alternatives to consider, each with its own investment requirements, time table, and projected stream of future costs and benefits. A second basic question is to determine which project – which projected set of cash flows – is most desirable. While these are not the only questions that must be answered to determine whether a project can be justified, they are questions that will certainly be considered by investors, bankers, and entrepreneurs who are considering participating in such projects. It is therefore essential to understand how such financial comparisons are made.

The first critical concept concerns the **time value of money**. There are several reasons why money in the future is worth less than the same amount of money available in the present. First, if money is available today, there is an **opportunity cost**: the money could be put into a savings account or invested so that it will be worth more in the future. Second, **inflation** is likely to reduce the purchasing power of money, so that today's money will buy more goods and services than the same amount of money would be expected to purchase in the future. Third, there is a **risk** that the actual money that becomes available in the future will be less than was predicted. It is therefore necessary to **discount** future cash flows in order to determine their **present value**. The higher the discount rate that is used, the lower the present value. The greater the potential for growth, the higher the expected rate of inflation, and the greater the risk associated with the proposed investment, the higher the discount rate that will be used.

By discounting costs and benefits over the life of a project, it is possible to determine the **net present value (NPV)** of the project. If the NPV is positive, then the project is worth more than doing nothing; if the NPV is negative, then the project is not worth pursuing, at least from a financial perspective. By estimating the NPV for a set of alternatives, then it is possible to determine which is worth the most financially. Maximizing the net present value of cash flows is a common financial objective for the private sector, although public projects are likely to have more complex objectives.

Discounting provides a means of establishing **equivalence** between two sets of cash flows. If the cash flows are equivalent for a company or an individual, then that company or individual is indifferent between them. The conceptual power of discounting is that it provides a straight-forward methodology for establishing a NPV that is equivalent to any arbitrary set of projected cash flows. In particular, it is possible to establish equivalence between a **present value P** and a **future value F** or an **annuity A**. Factors that relate P, F, and A are functions of the discount rate and time. These equivalence factors make it possible to compare cash flows in various ways, depending upon what is most useful for a particular analysis.

Equivalence relationships are used to determine **mortgage** payments, the **value of bonds**, and other financing mechanisms. A mortgage is a financing mechanism in which a bank or other financial company provides a large payment in return for a series of monthly payments over a period of many years. If the mortgage payments are not made, then the mortgage is in default, and the bank can foreclose on the property. A company or agency can sell bonds to raise money for a project; the bonds are purchased by investors who receive interest payments over the life of the bond and also receive the full value of the bond at the end of life of the bond. If the company or agency fails to make the interest payments or is unable to redeem the bonds as they become due, then the company or agency may be forced into bankruptcy.

The concept of equivalence can be used in cost models, where it is often necessary to consider both investment costs and annual operating costs. The investment cost must be converted to an equivalent annual cost in order to be added to operating cost and used to determine such things as the cost per unit of capacity or the cost per user.

Choosing a Discount Rate

The world's largest iron-ore producer [sold] $1 billion in investment grade bonds due in 2016 priced to yield 6.254%. Cia. Vale do Rio Doce, which last year became the first Brazilian corporation to win an investment grade rating, issued the debt to help fund its repurchase of $300 million of its 9% bonds due in 2013, thereby cutting the company's borrowing costs.

CVRD Issues Record Bond, LatinFinance, February 2006, p. 4

Introduction

The concept of equivalence is essential to evaluating infrastructure projects that require substantial investments before any benefits are obtained. Given a discount rate, it is possible to relate any arbitrary sequence of cash flows to an equivalent present worth or future worth or to an annuity that continues indefinitely or for a fixed number of periods. Given the projected costs and benefits, it is possible to calculate the net present value of a project, which can then easily be compared to the net present value for other projects. However, this extremely useful concept depends upon having a discount rate, and the selection of a discount rate is a complicated and ultimately quite subjective matter. The discount rate cannot be established by fiat, nor is there a methodology that can be used to determine the exact discount rate that someone should use in evaluating a project. Moreover, the various people and organizations considering a project are likely to use quite different discount rates in evaluating the same project. Since the discount rate determines the importance of future financial costs and benefits, it is necessary to give some thought to the choice of a discount rate.

The discount rate is similar, but not identical to the rate of return. The discount rate is a conceptual figure that is useful in establishing equivalence of cash flows, whereas the rate of return is an accounting term that is used in describing past or predicted financial performance of companies. Historically, rates of return have been higher for riskier investments, because investors discount future earnings more heavily for such investments. The greater the perceived risks associated with stocks or bonds, the lower the prices that they will command – and the higher the costs for the company or agency trying to raise capital for a project. The discount rate used to evaluate projects should be at least as high as the rate of return that could be obtained via other investments with similar risks. The **minimum attractive rate of return** for a company will be at least equal to its cost of capital – and perhaps much higher.

The concept of a discount rate may seem to be a rather arcane topic, but it has clear and important consequences for a company or a government agency. The quote at the beginning of this chapter describes how a company was able to reduce its interest costs from 9% to 6.254% because the financial community upgraded the company's credit rating. With lower interest on $300 million worth of bonds, the company saved approximately $8 million per year in interest from 2006 through 2013. How the financial community views the risks associated with a company can be extremely important to the profitability and even the survival of that company.

Profits and Rate of Return vs. Net Present Value

Companies and investors often think in terms of profits and return on investment as well as or instead of present worth or future worth. Profit and return on investment are both accounting terms. Profit is the difference between revenue and expense, while return on investment is the ratio of profit to the total amount invested. In the simplest case, an investment of I at time 0 results in annual profits of A/year over a very long time horizon. In this simple case, the annual return on investment would always be A/I. While this simple logic is fine for, say, buying bonds, it is insufficient for investments in infrastructure. The first problem is that the investment does not occur at time 0, but may in fact require years of effort. The second problem is that the revenues and expenses associated with the project are likely to vary over the life of the project. The third problem is that tax laws and accounting conventions determine what is called an expense and what is called an investment; the way that legislators and accountants consider financial matters may be quite different from the ways that entrepreneurs, companies and investors do their analysis.

The first two problems can be addressed by use of the concept of equivalence, assuming that an appropriate discount rate is given. The actual investments that take place over a period of months or years can be related to an equivalent total investment either at time 0 or at the time the project is completed and the operation begins. Likewise, the net benefits can be converted to an equivalent long-term annuity that begins either at time 0 or at the time that operation begins. The return on investment would then be the ratio of the equivalent annuity to the equivalent investment.

The problems related to accounting and taxes are trickier. Profit and return on investment are defined by legislation and accounting rules, and it is not possible to adjust the definitions of these terms to be consistent with some other view of the world, e.g. the emphasis on net present value of cash flows that is presented in this text and other texts on engineering economy, management science, and project evaluation. The generally accepted belief is that managers and investors will do better by focusing on the net present, future or annuity value of after-tax cash flows rather than focusing on profits as defined by accounting rules and government regulations.

It will eventually be necessary to consider how taxes and accounting rules affect cash flows and to understand how changes in tax laws and accounting can be used to promote different types of projects. For now, however, we keep our focus on cash flows without worrying about taxes. But we do have to worry about risk, and the easiest way to understand why is to consider the effect of borrowing money on the chances for a project's success, as demonstrated in the next section.

Leveraging and Risk

Borrowing the funds needed for an investment will reduce the initial investment required from the owner and potentially increase the expected return on the investment, but also increase the risk of the project. Because of the higher risk associated with the project, a higher discount rate will have to be used.

Consider the case from the previous section in which a project's cash flow has been transformed into the equivalent cash flows represented by an investment of I at time 0 and annual profits A that are received at the end of every year thereafter. If all the investment funds are provided by the owner of the project, then the owner's return on investment will be A/I. If the owner borrows a portion of the investment, then two things happen: the owner's investment declines by the amount borrowed, and the annual profit declines by the amount paid in interest. The owner's return on investment will now be:

(Eq. 1)   Owner's ROI = (A – loan interest)/(I – loan principal)

Figure 1 illustrates what would happen in a situation where the initial annual return A is $20 million on an investment I of $100 million. The y-axis on this chart should be interpreted as millions of dollars for income and debt payments, but for ROI it should be read as the annual %. The chart on the left shows the initial situation in which all of the investment is provided by the owner; the annual income is $20 million, there is no interest payment on the debt, and the ROI is 20% (A/I = $20 million/$100 million).

The chart on the right shows what happens if the owner only provides half of the investment and borrows the other $50 million at an interest rate of 10%. In this case, the annual income remains unchanged at $20 million, but there is an interest payment of $5 million. The owner's ROI jumps to 30%, as the net income to the owner after paying the interest is $15 million, which is 30% of the owner's investment.

Figure 1 Benefits of Leveraging: debt financing increases expected returns if the interest rate is lower than the non-leveraged ROI

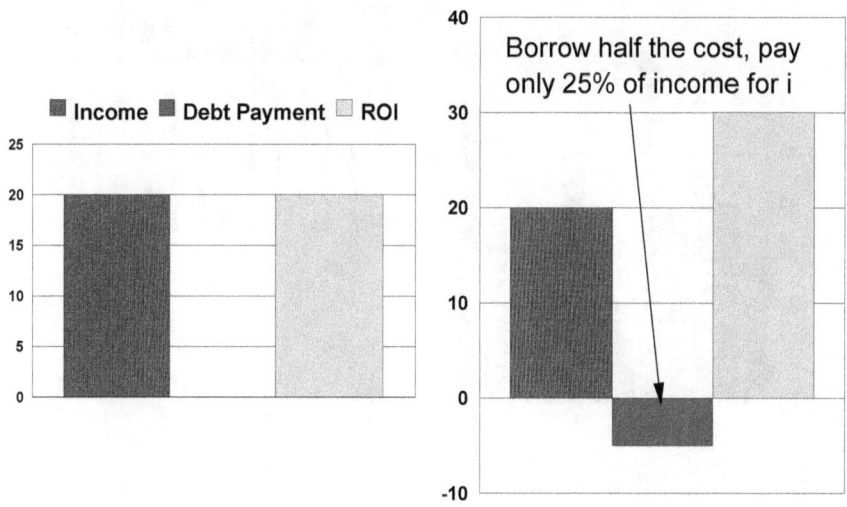

Borrowing money will increase the owner's ROI so long as the interest rate is lower than the return on the original project. The process of borrowing money in order to increase the amount that can be invested is called **leveraging**. Leveraging may increase or decrease the profitability of the project. Since infrastructure projects generally require very substantial investments, most such projects are highly leveraged. Minimizing interest costs becomes a major concern in such projects.

Whether leveraging is undertaken in order to increase ROI or simply to enable the project to be constructed, leveraging will increase risks because of the possibility of a default on interest payments. Figure 2 shows why this is so. As suggested by the chart on the left, cash flows may vary from year to year. In this situation, annual profits range between $5 million and $70 million. In the unleveraged situation, smaller annual profits simply reduce the profitability for that year, and the owner still enjoys positive cash flow. In the chart on the right, it is assumed that the project is very highly leveraged, requiring interest payments of $35 million per year. As a result, the cash flows in some years are insufficient to cover the interest payments, and the owner must use other sources of funds to make the interest payments and avoid bankruptcy.

The greater the uncertainty in predicting future cash flows, the greater the risks from leveraging. Leveraging works best in situations where there is a stable source of revenue, so that there is little risk that interest expenses will exceed actual cash flows. A project whose financing is heavily leveraged will be riskier than if it had little or no debt, and investors will therefore discount projected cash flows using a higher discount rate.

Figure 2 Risks of Debt Financing: leveraging increases risks because principal and interest payments must be paid when due

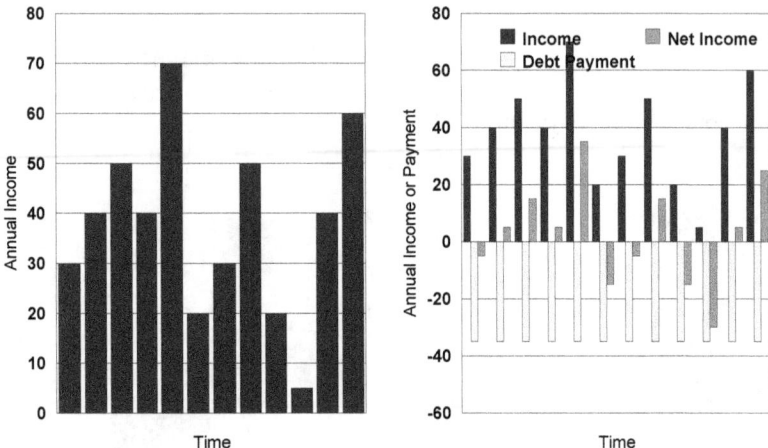

Factors Affecting the Discount Rate

Minimum Attractive Rate of Return (MARR)

The discount rate reflects the time value of money to a particular individual or organization. Future cash flows must be discounted because of at least three factors.

- **Opportunity cost**: the opportunity cost represents the potential financial benefits that must be foregone once a decision is made to invest in a particular project. Instead of investing in a particular project, individuals and companies could invest in other projects, in stocks and bonds, in real estate or in other ventures. If they have borrowed money, they could pay off their debt. If they have sold stocks to the public, they could buy back some of the stocks.
- **Inflation**: inflation is likely to erode the purchasing power of money received in the future. As a result, the same amount of money would purchase less in the future than it would today.
- **Risk**: the money that is anticipated to be received sometime in the future may or may not materialize.

These factors overlap to some extent, as the opportunity cost depends to some extent on expectations concerning inflation and risk. Also, each of these factors will be treated differently by those promoting a project and those investing in a project. Nevertheless, both promoters and investors are likely to have a **minimum attractive rate of return (MARR)** as they contemplate undertaking or investing in a project. The MARR represents the rate of return that promoters or investors believe that they could achieve via other investments with similar perceived risks. Discounting cash flows using the MARR as the discount rate will indicate whether or not the proposed project is as attractive as pursuing the other options that are available. Promoters and potential investors have different investment options and they have different perceptions of the risks associated with a particular project. Therefore the MARR used by those proposing a project is likely to be different from the MARR used by those who are asked to invest in the project. The differences are worth discussing in further detail.

MARR for Investors

Potential investors have quite a different perspective from those proposing a project. They are not necessarily concerned very much or at all with the objectives of the project, as they are primarily interested in maximizing their financial returns. They probably do not understand nearly as much about the technologies, locations, opportunities, or possibilities as do the promoters of a project - and the investors are apt to be leery of promoters who perhaps have reasons to oversell their project proposals.

Investors have numerous investment opportunities, and they have access to many qualified investment analysts who rate the financial attractiveness of many of those investment opportunities. They are likely to have a very good understanding of the financial markets, and they will have their own track record with respect to investing in different sectors and in different types of securities.

In general, investors can seek higher returns by investing in riskier endeavors, as illustrated in Figure 3. Savings accounts insured by the U.S. government and bonds issued by the U.S. government may be viewed as a risk-free investment, and they may offer an interest rate of 5% or less. Riskier bonds will require higher interest rates in order to attract investors. The risk that a bond will default can be estimated by analyzing the finances of the company or government agency issuing the bonds. Companies such as Moody's rate bonds with respect to their credit-worthiness; the best bonds are rated as AAA, then AA, B, etc. Bonds with lower ratings require higher interest rates because the probability of default is greater; bonds with very low ratings may be deemed unsuitable as investments for some very conservative pension funds or mutual funds. Bond prices reflect the discounted value of the stream of interest payments plus the discounted value of the ultimate redemption of the face value of the bond. The lower the credit rating, the more the interest and the ultimate redemption of the bond will be discounted. Since bonds commonly are sold in standard denominations (e.g. $1,000), the interest rate must be increased for riskier bonds to justify the sales price of $1,000.

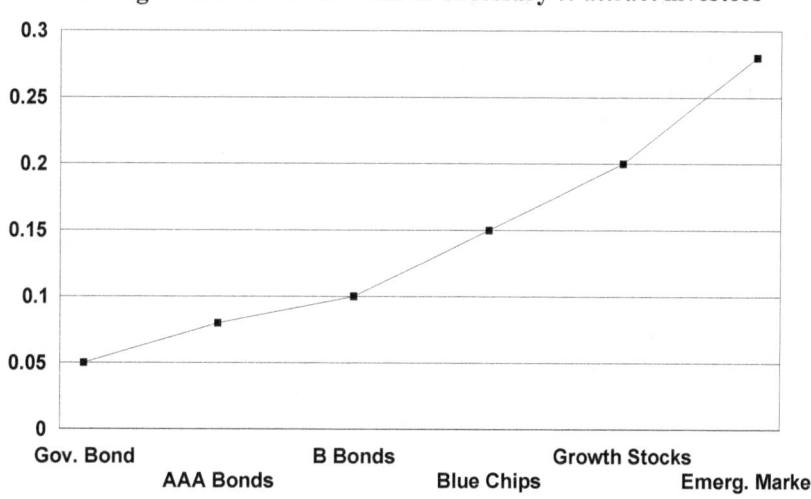

Figure 3 Risk vs. Expected Return: the higher the perceived risk, the higher the return that will be necessary to attract investors

Figure 3 shows stocks as having a higher expected return than B-rated bonds. This makes sense, because the interest must be paid on the bonds before any cash is available to the stockholders. "Blue Chips" are stocks in the largest, most well-known, and very financially stable and attractive companies. Blue Chips have an excellent history of good financial performance, and they therefore are viewed as having relatively low risk. The value of Blue Chip stocks is largely dependent upon a continuation of past performance rather than anticipation of future improvement. Growth stocks are issued by companies that anticipate rapid growth in earnings; the value of the stocks is based more on hope for the future rather than continuation of the past. Since the earnings are expected, not proven, investors discount future cash flows more heavily than they discount Blue Chip cash flows. Even higher discounts will be applied to cash flows projected by companies involved with new technologies or companies that do substantial business in regions of the world where financial markets are relatively undeveloped. The opportunities may well be much greater in such markets, but the risks are also much higher.

The risk-return curve plotted above in Figure 3 is determined by market forces and economic conditions. The shape of the curve is affected by the state of the economy, government policy concerning interest rates, government debt levels, the number of new offerings of stocks and bonds, political uncertainty in various regions around the world, public perceptions concerning new technologies, wars, and other factors. The discount rates implied by the risk-return curve can be interpreted as resulting from the sum of three factors:

- The real return available on risk-free investments
- The expected rate of inflation
- A risk premium that reflects investors' collective views concerning the riskiness of the company, including judgments concerning the company's financial situation, the outlook for the entire industry, and the social and political conditions in the country or countries where the company is located or where the company's products are sold.

Each of these factors is used to discount the current value of future earnings. For example, given the risk free return (rf), the annual risk premium, and the inflation rate (inf), the present worth of the cash flow CF in month N would be:

(Eq. 2) $\quad$ Present worth of $CF(N) = CF(N)/((1+rf)^N(1+risk)^N(1+inf)^N)$

If each of the three factors is small, then

(Eq. 3) $\quad$ Present worth of $CF(N) \sim CF(N)/(1 + rf + risk + inf)^N$

The discount rate, in this case, would be (rf + risk + inf).

It is important to understand what the chart in Figure 3 does show and what it does not show. It *does* show that investors will discount future cash flows much more severely if they believe that there is a lot of risk associated with a project. It does *not* show that investments in riskier projects will make more money. It *does* show that reducing risks allows a developer or a company to raise more money *today* based upon its projected cash flows for *tomorrow*.

It is even more important to understand that the above chart represents the market's evaluation of risks and investment opportunities. This is not the same as the owner's or developer's assessment of opportunities. Each individual and each company has their own Minimal Acceptable Rate of Return (MARR) and their own preferences for risk and return. Their MARR is based upon their opportunities, perceptions of their risks, and their preferences for risk and return. They will discount projections of future cash flows in their projects using their MARR for projects they deem to have similar risks. Because of their own experience or knowledge or ignorance, they may well believe that a project can be completed as planned and that it will indeed achieve the anticipated cash flows. Hence, they may see projects as being less risky (and therefore a better alternative) than the same projects would be viewed by the market. They will therefore go to rather great lengths to convince potential investors that the project is feasible and that the risks are not as great as the investors might fear.

MARR for Owners and Promoters

Project owners and promoters have a much different perspective than investors. The promoters may be a developer, a company, or an agency promoting a particular project; there may be a "champion" who has long been advocating the project and seeking political and financial support from many agencies, organizations, and companies. The promoters are seeking to achieve various benefits from the project, some of which may be financial, some of which may be economic and quantifiable as monetary benefits, and some of which may be social or environmental or other benefits that are difficult to quantify. They understand that they need to finance the project, i.e. come up with the cash necessary to pay for the investment and the continuing expenses of the project. Presumably they believe that the project is worth doing, that it is a good way to achieve the benefits that are sought, and that it has a reasonable chance of success. They will have considered similar projects and perhaps they may have already completed similar projects. They likely have consulted experts regarding the feasibility of the construction, the demand for the project and the risks associated with the proposed project. They also have a good idea of their own opportunity costs, based upon their experience and their investment options. They presumably do not expect their next project to be their best project ever, but they do expect to achieve returns similar to what they have achieved in the past and to what they believe they could achieve in other projects.

If the promoters have sufficient funds to construct and operate the project, then they can do pretty much what they want to do. If they need to raise money for the project, then they have additional financial concerns. If they borrow money, they will worry about the interest rates they will have to pay. If they need to sell stock, they will worry about how what price their stock will command in the market and what proportion of the ownership must be transferred to the new stockholders. Their MARR must be greater than their cost of capital and it must be greater than what they earn in other endeavors.

The cost of capital is an important factor for companies. Many companies raise money by selling bonds or selling stock. The cost of capital is lower for bonds than for stocks, as described in the previous section. For bonds, the cost of capital is the interest rate that is paid. For stocks, the cost of capital can be estimated as the historical returns to the owners of the stock. The annual return for a stock can be calculated as follows:

(Eq. 4) Annual return = (Final Value − Initial Value + Dividends)/Initial Value

For example, consider a stock that was valued at $100/share on January 1st, paid a $2 dividend on December 31st and had a value of $110 on January 1st of the following year. The stock increased in value by 10% and the dividend was 2% of the initial value, so the total return was 12%. The average historical returns for a stock can be used when estimating the cost of capital for that company.

A share of stock represents ownership in the company, which means that stockholders have "equity" in the company. The **weighted average cost of capital** depends on the relative amounts of debt and equity in the company and can be calculated as follows.

(Eq. 5) WACC = %D (average interest on debt) + % E (average return on equity)

Where

%D = Debt as percent of total value of debt plus stock

%E = Equity as percent of total value of debt plus stock

For example, we can calculate the weighted average cost of capital for a company whose ratio of debt to equity is 3::2, whose average interest rate on debt is 8%, and whose historical returns to equity have been 16%. Substituting these values in Equation 6 shows that the company's average cost of capital is 11.2%:

(Eq. 6) WACC = 60% debt (8% interest on debt) + 40% equity (16% return on equity)

= .6 (8%) + .4 (16%) = 4.8% +6.4% = 11.2%

As discussed above, a company may reduce its credit rating and increase the risk of bankruptcy if it increases its proportion of debt. If a company increases the ratio of debt to equity, it will eventually encounter higher interest rates and difficulties in selling bonds. To maintain their credit ratings, many companies will endeavor to maintain a constant ratio of debt to equity. If this ratio is constant, then the WACC represents what it costs (or has cost) the company to raise capital for new projects. Over time, as the debt/equity ratio and market conditions change, the cost of capital will also change.

The analysis of the cost of capital can be made much more complicated. Instead of the weighted average cost of capital over the past year or past five years, a company will likely be interested in its current cost of capital. In regulatory affairs and legal cases, lawyers and experts will debate projected cash flows and discount rates. In Wall Street or other financial centers, analysts will be trying to figure out the true values of stock prices, which presumably reflect the projected cash flows for the company, the discount rate applied to those cash flows, and the number of shares outstanding. Company financial officers present their projections of cash flows to financial analysts, and the financial analysts make recommendations to their clients as to whether the stock should be bought, sold, or held. The financial analysts may make their own judgments regarding future cash flows of the company, and they may select a discount rate based upon their perceptions of risk. Their judgments will be critical in determining share prices as well as the interest rates the company will have to pay on future sales of bonds.

The benefit to a company in issuing stock instead of bonds is that there is no requirement to issue dividends, which is very helpful during unprofitable periods. If a company defaults on interest payments on bonds, it can be forced into bankruptcy, but if it decides not to issue dividends, there is no such risk. The negative aspect of selling stock is that the sales price may reflect investors' application of very high discount rates to the company's projection of future cash flows. As a result, the current owners may give up more of the company (and its future earnings) than they want to if they try to sell more shares of stock.

It is important to emphasize again and again that perceptions of risk and choices of discount rates will be different for the different parties involved in a project. For example, suppose a group of engineers and entrepreneurs has created a company that is trying to raise $100 million for constructing what they believe to be a very lucrative project in a developing country. The company has done extensive research concerning construction costs, operating cost, and potential revenues. While they recognize that there are some risks related to regional economic conditions, they are confident that their proposal can be constructed on time and on budget and that it will attract high demand almost immediately. They have prepared a prospectus in which they describe the project in great detail, and they anticipate annual profits to reach $20 million per year for at least 30 years. Using a discount rate of 10%, they estimated the net present value of the revenue stream to be $200 million as of the time the project begins operations, which is far greater than the construction cost. They will have no income for several years while the project is being constructed, so they hope to raise the $100 million from a combination of low-interest loans and sale of stock to get started. Almost immediately, they discover that they would have to pay 12% interest for a construction loan of no more than $20 million; by the time construction was completed, they estimated they would have to pay $5 million in interest. Then they approached several potential investors to see if would be feasible to sell stock in their company to raise the funds necessary to cover the remaining $80 million of construction costs plus the $5 million in interest. They indeed found investors interested in infrastructure projects in this country – but those investors were discounting cash flows by 25% or more when they evaluated such projects. Hence, for those investors, the cash flows of $20 million per year were worth only $80 million (estimated as the equivalent present value of an annuity for an indefinite period = $20 million per year/(0.25) = $80 million). Thus, to these investors, the returns from the proposed project were much less than their MARR, so that they would not be willing to pay enough for the stock for the developers to finance their project even if they purchased 100% of the company. The project therefore would likely collapse, because it would be impossible to raise sufficient funds to get started.

MARR for Large Companies

A large company's minimum attractive rate of return (MARR) will never be less than its weighted average cost of capital (WACC), because it can always buy back its stock and bonds if it has an excess of cash and no suitable investments. Since the WACC reflects market valuations of its stock and bonds, the WACC and therefore the MARR will be affected by macro-economic factors such as inflation and recessions as well as by the market's perception of risks associated with the company and the industry.

The company's MARR must also take into consideration the company's opportunities for investments. Investments in the company may be desirable in order to expand capacity, increase efficiency, support moves into new types of business, or to improve safety. Company officials will have many competing internal requests for capital, and they will have (or should have) an excellent perspective concerning the potential benefits for these competing projects. A company may also consider broader investments. It can invest in the financial markets, just like any investor, or it could attempt to buy or merge with other companies.

In a completely predictable world, a company perhaps would be able to invest in any and all projects with a return greater than its WACC. After all, if a company can raise money by selling stocks and bonds at an average cost of, say 11%, then it can increase its profits by investing in all projects with an expected return greater than 11%. If so, then the MARR for this company would be the WACC.

In actual practice, however, the company is faced with many uncertainties. The projections of cash flows are based upon numerous assumptions, and there is in fact a potential distribution of cash flows associated with each project. The expected return may be greater than 11%, but there may be a good chance of earning less or even substantially less than 11%. The more highly leveraged the company becomes, the greater the potential that it will be unable to cover its interest payments. The company may not be willing to risk losing money, and therefore may not wish to invest in projects unless they are very sure that the projects will have a return greater and perhaps substantially greater than its WACC. This kind of financial discipline will be imposed at three levels:

- The Chief Financial Officer will scrutinize proposals and impose criteria for ranking and selecting projects, taking into account the need to maintain a debt/equity ratio that is acceptable to investors
- The Board of Directors may limit the total capital budget for the company in order to limit the overall risks associated with the company
- The capital markets may decide that a company is a credit risk, which means that someone will issue a report or a study or a credit ranking that reflects poorly on the company. Companies, countries or agencies with a poor credit rating will have to pay higher interest on any future bonds that they issue and they will probably see a reduction in their stock prices or more difficulty in raising money for their projects.

Thus, in practice, a company will typically be unable to invest in all of the potentially profitable projects. Instead, it is likely to invest in the most profitable projects, which means that it will set its MARR (often called a "hurdle rate") well above its average cost of capital.

MARR for State, Local, and Federal Governments

Governments are often able to sell bonds at very low interest rates, so it may appear that the cost of capital for a public agency is very low. However, the low rates reflect in part the ability of governments to raise money by taxation. The actual cost of capital reflects not only the interest costs on public bonds but also the opportunity cost associated with taxation. What could the public have otherwise done with the money they paid in taxes? When economists consider this question, they come up with a higher cost of capital and therefore a higher discount rate or MARR for public projects.

Consider a taxpayer who has a minimum acceptable rate of return of 12%. This individual or business would be unwilling to invest in a scheme that offered less than 12% profit. The taxpayer would not keep all of the profits, as any applicable federal, state, or local taxes would have to be paid. If those taxes amounted to a third of the profits, then the minimum acceptable after-tax profit for the taxpayer would be 8% of the investment. A taxpayer concerned with efficiency in government might feel that government funded projects should be subjected to the same kind of test: the government should not invest in projects unless they also provide an after-tax benefit of 8%. Since governments do not pay taxes, this taxpayer might conclude that the government should use a minimum acceptable rate of return of 8% for projects funded by general tax revenues. The taxpayer might prefer projects to have a financial return of 8%, but perhaps would be willing to include clearly documented economic benefits received by society.

There is certainly some credibility to the argument that governments should not use tax revenues for projects with only modest benefits if the same money could have been better used by the taxpayers themselves. Deciding what rate of return is required is a matter of policy more than logic. In the United States, the General Accountability Office from time to time issues guidelines for the discount rates that should be used for public projects. These rates have recently been 7-8%, which is consistent with the logic expressed above. Given the many difficulties in measuring and monetarizing the public costs and benefits of a project, the choice of a discount rate is just one of many analytical assumptions that must be considered in evaluating a public project.

MARR for Special Government Agencies

Governments sometimes create special independent agencies that are authorized to raise money by selling bonds. These bonds are backed not by taxation, but by the financial credit of the agency. For these agencies, the cost of capital really is the interest rates that they must pay on their bonds. These agencies are therefore able to invest in projects with low financial returns – presumably justified by the broader socio-economic benefits provided by the agency. Examples would include port authorities, turnpike authorities, and housing authorities.

Choosing a Discount Rates: Examples

This section presents several examples that illustrate the logic behind the choice of a discount rate. Note that the choice is heavily dependent upon the perspective of the individual or organization that is making the choice. In some situations, there may be no clear answer, and there is usually no precise answer.

Determining the Interest Rate for Corporate Bonds

As the VP Finance for Acme Construction, you have been asked to estimate the interest rates on bonds that your company plans to sell in order to finance the construction of a new toll bridge. Long-term US Treasury Bonds currently pay just under 5% interest, and you know that investors usually consider these to be nearly risk-free investments. You expect that the rate of inflation (currently 2%) will increase to 3% by the time that your company is ready to sell the bonds. You expect that your company's bonds will continue to have a risk premium of 2% relative to US Treasury bonds. What interest rate should you plan to pay on these bonds?

This question addresses the interest rates that investors will require to invest in your project. From a financial analyst's perspective, the interest rate will be determined by the market. As seen in section 8.3, the market requires higher returns for riskier securities (see the figure that illustrates a risk/return curve). Discount rates for each type of security will equal the sum of three factors:

- the interest rate on a risk-free, inflation-free investment (i.e. the basic time value of money)
- the inflation rate
- a risk premium

Here, we have 5% interest on US Treasury bonds, which includes inflation, but is presumed to be risk free. We expect inflation to increase by 1%. And the risk premium will be 2%. Therefore we expect the rate to be 5%+1%+2% = 8%

Choosing a Discount Rate Based Upon a Firm's Cost of Capital

Brothers K, a construction firm specializing in prison security, borrows money at 6%; its stock is priced by the market to provide a 12% return. The company's debt/equity ratio is currently 2. What discount rate should the company use to evaluate new projects so as to at least cover its weighted average cost of capital?

We know that the company's MARR should be at least as great as the weighted average cost of capital. We are given no further information about the company, except that it is an established firm with prior projects and an ability to raise money by borrowing or by selling stock. The debt/equity ratio of 2 means that loans account for 2/3 of the market valuation of the company, while stocks account for 1/3 of the market valuation. The weighted average cost of capital will be 2/3(6%) plus 1/3(12%) = 8%. It would be possible to argue for a higher rate based upon the likelihood that the MARR will be above, or even significantly above the WACC.

Choosing a Discount Rate for a City

Cities must justify investment in infrastructure by comparing the present worth of net benefits over the life of the project to the present worth of the construction costs. Cities raise money through income tax, property tax, or the sale of bonds. Interest paid on municipal bonds is about 4%, which is lower than the interest rates that private companies pay on their bonds (about 6%) and much lower than the returns required to sell stock (10% or more). What discount rate should the city use in evaluating the present worth of their investments?

A city's discount rate must be greater than the interest it pays on bonds (4%) because it raises money from taxpayers – individuals and companies - who have other options. The cost of capital for taxpayers will reflect some mixture of debt and equity financing, i.e. about 8% if the Brothers K firm in the previous example is typical. As noted above, the General Accountability Office requires the U.S. federal government to use something like 7-8% for discounting related to public investments.

Choosing a Discount Rate Based Upon a Firm's Investment Options

Earp Enterprises is a developer in Arizona that specializes in building relatively inexpensive, but functional corrals which they call "OK Corrals". Because of the company's use of advanced planning techniques and standardized components, it has consistently been able to make a return of 14-18% on its projects. The company's weighted average cost of capital is 10%, and it is a highly profitable company. A recent PhD from MIT, Flora Holliday, would like Mr. Earp to invest in a second home development focused on a new marina at Lake Powell. She has prepared a business plan that shows the cash flows that she expects from this venture, which she calls "Holliday Docks". What discount rate should Earp Enterprises use to determine whether or not to shoot down this proposal?

Earp's MARR must be greater than or equal to his cost of capital, which is 10%. If capital is unlimited, then any project that earns the cost of capital will be acceptable. On the other hand, capital probably is not unlimited and the question certainly implies that Earp has done very well on his prior projects. Since Earp has won many financial victories at his OK Corrals, he seems to have opportunities to make 14-18% "consistently". Hence, his MARR should reflect the lucrative potential for continued investment in OK Corrals and be higher than the WACC, perhaps 14-15%.

Dividing Up the Cash Flows of a Major Project

It is important to understand that the various players involved in a major project will have markedly different cash flows, different risks to worry about, and different MARRs. Consider a case where a developer has secured a line of credit from Bank One for constructing a building. The bank will pay all the construction costs and charge the owner interest; no payments will be made on the loan until the project is completed, at which time loan payments will begin. The construction loan is likely to have a high interest rate, since there are risks related to the feasibility, time, and cost of construction. The owner plans to refinance the loan with Bank Two when the building is completed at the end of year 3. Refinancing should provide a lower interest rate because the building will in fact be completed and there will (hopefully) be tenants who are paying on long-term leases. If all goes well, the new loan will cover the construction loan and the monthly loan payments plus operating costs will be less than the revenue from the tenants.

The cash flows for the three major players will be as follows:

- Bank One – pays all construction costs as they are incurred over the three years of construction; receives reimbursement plus interest when the loan is refinanced at the end of year 3.
- Bank Two – gives the owner an amount large enough to pay off the construction loan at the end of year 3; receives monthly payments of principal and interest for the life of the mortgage.
- Owner – pays nothing during construction period, since all of those costs are covered by Bank One; receives a large amount from Bank Two at the end of 3 years but immediately uses that money to pay off the loan from Bank One; collects lease payments, pays for operations, and makes loan payments to Bank Two over the life of the loan.

Bank One has completed its role by the end of year 3. Bank Two just starts its role at that time. The owner, if all goes well, doesn't have to put up his own money to construct the building, and then has sufficient cash flow to cover the mortgage payments. Bank One prefers to earn a high rate of interest rather than to hold on to its cash; the owner prefers to have the cash as needed. Bank Two is willing to accept lower interest, but is also creating a long-term annuity for itself. The owner would rather pay the interest on a long-term loan than pay for the building when it is constructed.

We could go into more detail and consider such things as the cash flow for the construction firms and suppliers or the possibility of selling the building upon completion. Each actor will have different perspectives on whether or not this is a good project, and each will have a different level of exposure to the risks that might be associated with the project.

Summary

Discounting is a mechanism for converting an arbitrary stream of cash flows into a present value, a future value, or an annuity. Three main factors must be considered in choosing a reasonable discount rate:

- Investment opportunities: what alternative opportunities are available for investment?
- Risk: is the proposed project more or less risky than the other options?
- Inflation: how much will inflation reduce the future purchasing power of of our money?

Discount rates and the notion of a "Minimum Acceptable Rate of Return" (MARR) are very important and potentially very confusing topics. The confusion results because of the differences in perspective among the various actors involved in designing, building, and financing a project.

Developers and entrepreneurs are generally in the position of raising money for constructing projects that they believe to be justifiable in terms of their future benefits. Since they lack the funds to build these projects, they must convince others to invest in them, possibly by citing the importance of the project, but more likely by demonstrating the potential profitability of the investment. Developers have various strategies for raising funds for their projects, including borrowing money from a bank, selling bonds, and selling stock.

Investors have a much different perspective than developers. Developers are thinking of receiving rents or tolls or operating profits, which they hope will be enough to cover the mortgage or interest payments or to justify a high price for their stock. For them, the cost of money is similar to the cost of energy or labor, and their main concern is about the long-term success of their projects. For the bankers and other investors, the nature of the project is much less relevant than the prospect of making money from mortgage payments, loan payments, bond-interest or rising values for the stock they have purchased.

Financial markets exist for stocks, bonds, mortgages, mutual funds, and other types of financial assets. The price of these financial assets depends upon the market, i.e. upon the price that a willing seller will accept from a buyer. The value of these assets (to an investor or a securities analyst) is based upon a projection of cash flows, an estimation of the risks associated with these cash flows, and the availability and price of other assets with similar levels of risks. Different potential investors may view the cash flows, the risks, and the alternative opportunities quite differently, which is part of the reason why securities are continuously bought and sold.

The different perspectives of financial analysts, governments, independent government agencies, and private companies result in different ways of determining their MARRs. Financial analysts are not particularly concerned about the merits of a company or a project; their job is to determine the risks associated with stocks, bonds or loans associated with a company or a project. A company with good credit can easily raise capital to invest in bad projects; a company with no past history may be unable to raise funds even for projects that may appear to the public to be highly desirable.

When local, state or federal government agencies discount the costs and benefits of proposed projects, their discount rate should reflect the average returns available to taxpayers, not the low interest rates on public bonds. However, if a special public agency raises money by selling bonds rather than through taxation, then it can use the interest rate on those bonds as its MARR.

For private companies, the MARR should be at least as great as their weighted average cost of capital, and it should be at least as high as the rate of return achievable on alternative projects or on investments in the financial markets. In practice, the financial risks associated with leveraging will generally result in an MARR well above the weighted average cost of capital.

In any situation, the discount rate is a rather fuzzy number, so it will be wise to consider a range of discount rates when evaluating a project.

In major projects, it is usually necessary to raise funds for construction from banks (loans) or financial markets (stocks and bonds). An entrepreneur or a company presents estimates of costs and benefits to banks and investors, who then evaluate the risks and choose discount rates consistent with their own MARRs. If a commitment is made to pay the banks before paying interest on bonds or dividends on stocks, then the banks' investments in the project are less risky than the investments made by those buying stocks and bonds. If the project is being undertaken by a government agency or a large company, it may be possible to get a low interest rate for loans based upon the agency's or company's credit rating rather than a rate based upon the riskiness of the project.

If you can reduce the perceived risks of your project, you can raise more money because investors will apply a lower discount rate to the same future benefits. In particular, a project that has been completed or that has very clear commitments for cash flows (leases for a building; approved tolls for a highway; approved public subsidies) will have lower risks than a new project with uncertain time to completion and no guaranteed source of income.

"Leveraging" is a term used when money is borrowed for your project and a pledge is made to a) repay the loan or b) turn the project over to the lender if loan payments are not made. Borrowing reduces the amount of their own money that the owners must put into the project and allows a chance for a greater return on their investment, but also creates greater financial risks.

Financial Assessment

I will gladly pay you Tuesday for a hamburger today.

J. Wellington Wimpy (Popeye's friend)

Introduction

No project is undertaken solely to make money, but money is a consideration in every project. Entrepreneurs are in the game primarily to make a lot of money, even though the projects they contemplate are at some level aimed at satisfying someone's needs or desires. Even if a project is contemplated solely for some marvelous cultural or aesthetic benefit, it will still be necessary to pay the carpenters and buy the lumber. We therefore must be prepared to deal with money, to understand why someone would be willing to invest in a project, and to understand how entrepreneurs and investors think about projects and about money.

Owners and entrepreneurs need funds up front in order to create their projects; they expect future profits to be sufficient to provide an attractive return on their investment. If they try to raise money from a bank or from investors, they must prepare a financial plan that shows how the project will generate sufficient cash flows to pay off the interest on the loans or bonds, while increasing the value of the company for stockholders. When trying to raise money, what the owners and entrepreneurs think the project is worth does not necessarily matter very much. What matters is what the bankers and other potential investors think the project is worth. If they perceive the project to be very risky, they will use a higher discount rate. If their portion of the project can be made less risky, then they will use a lower discount rate and be willing to invest more in the project. It is conceivable that projects that appear very profitable to the proponents may appear to be too risky to investors, who will therefore be unwilling to provide the funds needed.

In general, a project must satisfy three criteria to be worth pursuing:

- The benefits expected from the project must be greater than its costs.
- The project must be viewed as a good way to achieve these benefits, because there may be engineering or institutional alternatives that are as good or better.
- There must not be better ways to use the resources that would be devoted to this project; maybe it would be better to invest in housing than in transportation.

Basic methods of engineering economics can be used to assess competing projects based upon analysis of their projected cash flows and any economic impacts that can be expressed in monetary terms.

Maximizing Net Present Value

The net present value (NPV) of a project is the difference between the present value of the net benefits over the life of the project and the present value of the investment. The NPV of a project will depend upon the costs and benefits that are considered, the project life, and the discount rate. In general, the objective will be to maximize the net present value when evaluating alternative projects.

If the NPV is positive, then any equivalent annuity and any equivalent future worth will also be positive. If the NPV of one option is better than the NPV of another project, then any equivalent annuity A or future value FV will also be better for this option. Whichever measure is used, the ranking of any options will be the same. Depending upon the situation, it may make sense to focus on future worth or annuities rather than NPV, as shown by the following examples:

a. Planning for a major future event, such as replacing a bridge: the basic question is how much to allocate each year to a sinking fund so that the future value of that fund will be sufficient to pay for the bridge replacement.
b. Incorporating equipment costs in operating budgets: operating budgets can easily include weekly or monthly expenses. Converting the purchase price into an equivalent weekly or monthly cost is therefore a convenient way to allocate costs of equipment.
c. Construction of an office building: the critical time is likely to be the completion of construction, so it will be useful to estimate the future value of construction costs as of that time. It will then be useful to convert the FV of the construction cost into an annuity that could be used compared to anticipated annual rent payments and maintenance costs.
d. Investments aimed at improving the environment, where the benefits may be measurable but not in monetary terms. If it is not possible to monetarize the benefits, then it will be impossible or meaningless to talk about the NPV of such benefits. Instead, convert the investment cost into an equivalent annuity over the life of the project. That way, the comparison of alternatives can be based upon cost effectiveness by comparing the expected annual benefits of each alternative to its annual cost.

NPV analysis is widely used because it can translate the cash flows of complex projects into equivalent amounts that are very easy to understand and to compare, assuming that a reasonable discount rate is used and acknowledging the fact that different parties involved in a project may have different discount rates.

The choice of a discount rate will be extremely important in determining what kinds of projects are most appealing. If a very high discount rates is used, then the NPV will be based primarily upon what happens in the first 5-20 years of a project. Small projects with immediate benefits will look better with a higher discount rate, whereas large projects with benefits that extend far into the future may fare poorly. If a very low discount rate is used, the opposite will be true: future costs and benefits will be much more heavily weighted.

In planning for public projects, use of a low discount rate may promote undertaking very large-scale projects while ignoring very important current needs. On the other hand, use of a very high discount rate may prevent a company or a country from ever undertaking large-scale projects.

Importance of Project Life

Projects need to be evaluated over a reasonable project life. Several factors enter into the choice of a "reasonable" life:

- The economic life of the project: the period of time for which the project is expected to be in use.
- The period of time for which discounted cash flows are relevant to the analysis.
- Knowledge concerning any dramatic costs or benefits that might be expected in the distant future.

The economic life of a project can be much less than the physical life of the structures that are constructed. If a railroad is built to a mine, the railroad might be expected to last indefinitely so long as it is maintained properly. The facilities at the mine may also be constructed to standards that would ensure a life of 30-50 years. However, the ore may be gone after just 20 years, so that the economic life of both the railroad and the mine would be 20 years.

Unless very low discount rates are used, a 20 to 50 year life is usually sufficient for analysis. Because of discounting, the costs and benefits from more distant years will not add much to the NPV, so it will not be necessary to include them in the calculations.

Ignoring the out-years has been viewed by some as something very bad, as it means that the analysis would be ignoring the impacts of current decisions on unborn generations. Some have called for the use of a zero-discount rate so that the needs of future generations would be considered properly. In a financial analysis, however, it is foolish to talk of a zero-discount rate since in fact investors and the financial markets that provide the funding for projects do discount cash flows – and the amount of money that can be raised for projects depends upon their discount rates. Potential benefits that occur in the far distant future will not attract additional funding from the markets.

Of course there may be some merit to the argument that current projects may be damaging the environment, creating hazards or promoting serious financial problems in ways that will not be apparent for 20 or more years. If a 20-year life is used, then such problems may conveniently be overlooked. This problem can be dealt with by requiring additional considerations in the choice of the time period:

- Are benefits expected to continue to exceed costs for an indefinitely long period?
- Will the project need to be decommissioned at the end of its useful life, and is the cost of that decommissioning included at the end of the assumed project life?
- Are there potential catastrophic consequences that could be caused by the project beyond what is considered within the chosen period for the analysis?
- Are there extraordinary costs or benefits that can be expected after the proposed project life?

For projects where nothing unusual is expected in the distant future, the use of a 20-50 year project life will be long enough to capture the relevant costs and benefits associated with a project. For discount rates of 5% or more, the out-years will contribute very little to the analysis, and it will be rather meaningless to make projections further into the future. If the economic life of the project is less than 20 years, then a shorter life should be used. If there is reason to expect extraordinary costs or benefits that would be apparent only after a period of 20-50 years, then the project life should of course be extended. In normal circumstances, using a discount rate and limiting the life of the project should not be seen as somehow damaging to future generations – it is simply reflecting the reality of money and the principle of equivalence.

Does Discounting Ignore Future Catastrophes?

To answer this question, we need to define what is meant by a catastrophe and what the costs of a catastrophe might be. To provide some perspective, we can look at the more dismal side of history. There have been numerous instances where natural disasters – earthquakes, hurricanes, or tsunamis - have killed tens of thousands of people, and there have been outbreaks of disease that have killed that many people in a single year in many different cities. In 2010, a horrendously devastating earthquake destroyed much of Port Au Prince and other cities in Haiti, killed on the order of 100,000 people, and left more hundreds of thousands injured or homeless. In a few minutes, this earthquake caused double the amount of casualties suffered by US troops in all of this country's wars from Viet Nam to Afghanistan, and it caused 30 times the loss of human life suffered on 9/11. Wars with tens of thousands of casualties are commonplace in history, and the world wars of the 20th century killed tens of millions. Epidemics, often initiated as a result of natural disasters or warfare, can have the most devastating impacts on humanity. Millions of people died during the Great Influenza of the early 1920s, and the Black Plague reduced the population of Europe by a third during the 14th century. Diseases introduced by Europeans wiped out an even larger proportion of the native populations of North and South America during the 16th century.

Natural disasters, warfare and disease will unfortunately continue to afflict humanity with catastrophic consequences far into the future. Whether we are planners, engineers, political leaders, or private citizens, we should all be concerned with ways that we could limit the frequency or the consequences of such catastrophes. If discounting really does make it possible to ignore catastrophic events far in the future, then that would be a severe flaw in the methodologies commonly used in project evaluation. However, the financial, economic, and social costs of catastrophes can be so large that they cannot be ignored, even if the risks are small or far distant in time.

For example, consider the possibility of an epidemic that could break out in 50 years, taking the lives of 1 million people. Suppose that steps could be taken today that would reduce the expected fatalities by 90% or defer the epidemic for another 50 years. What would the benefits be, assuming a discount rate of 8%?

We can quantify the magnitude of such a disaster using an approach that various countries have adopted in managing risks associated with accidents, infrastructure failure, and natural disasters. This approach evaluates the cost effectiveness of risk reduction strategies by comparing the costs of a strategy to the expected reduction in fatalities. In the United States and in Europe, government safety regulations can be justified if the costs of improving safety are less than about $2.5 million per expected life saved.

If we use this approach, then an event that led to 1 million deaths would have a cost to society of 1 million deaths multiplied by $2.5 million per fatality for a total of $2.5 trillion (i.e. 2.5×10^{12}). This is an extremely large number. Even if this occurs 50 years in the future, the NPV of such a disaster is very large. With a discount rate of 8%, the NPV would be $2.5 trillion $(1/1.08)^{50}$ = $2.5 trillion (0.02132) = $53 billion, which is not an insubstantial sum of money! Thus efforts that could reduce the expected fatalities by 90% would be worth nearly $50 billion today.

If the disaster occurred 100 years in the future rather than 50, then the $2.5 trillion would be further discounted by another factor of 0.02131, and the NPV would be reduced from $53 to a bit more than $1 billion. Thus, reducing the magnitude of this catastrophe by 90% or deferring the epidemic for another 50 years would each have a NPV of approximately $50 billion. Discounting does not allow us to ignore future catastrophes; it provides a rational way to assess the cost effectiveness of strategies for preventing, preparing for, or dealing with potential catastrophes.

What would projects look like that had the effect of reducing the frequency or consequences of future disasters? For reducing the probability and severity of a pandemic, doctors can work to develop better drugs, public health officials can work to eliminate unhealthy slums and improve water supplies, and governments can stockpile emergency supplies of medicine and other supplies. To reduce the consequences of earthquakes and other natural disasters, governments can impose building codes that limit or require sturdier construction in dangerous areas, they can provide better communications and warning systems, and they can prepare for rapid response to natural disasters. A lot can be done in each of these areas for $50 billion!

Return on Investment and Internal Rate of Return

When reporting their financial results, companies produce reports that follow generally accepted accounting procedures to document profitability and return on investment (ROI). The ROI for a given year is calculated by dividing the company's annual profit by the company's net investment. By projecting a company's revenues and expenses into the future, financial analysts can predict future levels of profitability and ROI.

When evaluating investment opportunities, companies seek projects that will increase ROI for the company as a whole. To do so, a new project must, over time, provide a return on the new investment that is greater than the company's actual ROI. Thus, while an engineering economist might prefer using NPV analysis, senior management might prefer to know the expected ROI for a project. Some projects are very straightforward. If a new machine costs $100,000 and saves $15,000 per year in operating expense, then the ROI will be 15% per year. However, major infrastructure projects will never be so simple, because the initial investment may be spread out over several years, and the anticipated cash flows will likely vary from year to year.

It is of course possible to convert the initial investment into an equivalent investment I at time 0 and to convert the net revenues into an equivalent annuity A that will continue forever. Once this is done, the return on investment is readily seen to be A/I. This is a very useful concept, but the result depends upon the discount rate that is used, and the choice of a discount rate depends upon the perspective of the user. Another approach makes it is possible to estimate of ROI for a project without depending upon a pre-determined discount rate: simply find the discount rate that will make the NPV of the project's cash flows equal zero, in which case the ROI will equal the discount rate. This rate is known as the **internal rate of return (IRR).** The higher the IRR the better, and companies in the private sector commonly use this method to characterize the profitability of proposed projects.

Ranking projects by IRR only works for projects that are independent. If projects are mutually exclusive, then a smaller project with a high IRR may prevent a larger project with lower IRR but much greater total returns.

For example, suppose a company has identified four options for expanding their operations, all of which would use the same site. Various financial measures are given in Table 1 for each of these projects, and you are trying to decide if the annual net benefits are large enough to justify any of the investments. The annual benefits are expected to continue for a very long time, so the net present value of the benefits was estimated using the capital worth method, i.e. by dividing the annual benefits by 10%, which is your firm's minimum acceptable rate of return. The annual net benefits of $90,000 for Project A would therefore be worth $90,000/10% = $900,000. Since this is less than the investment cost, project A has a negative net present value and should not be pursued. For the other three projects, the present value of the benefits exceeds the investment cost, so that the NPV for each of these projects is greater than zero. Projects B, C, and D can therefore all be justified financially. If only one of the projects can be undertaken, then C is the best, as it has the highest NPV.

Table 1 Investment and Benefit Data for Four Projects

Project	Investment (NPV as of time zero)	Equivalent Annual Net Benefits	Present Value of Benefits (Using capital worth method)	Net Present Value of Project	IRR
A	$1 million	$90,000	$0.9 million	($0.1 Million)	9%
B	$2 million	$440,000	$4.4 million	$2.4 million	22%
C	$3 million	$600,000	$6.0 million	$3 million	20%
D	$4 million	$480,000	$4.8 million	$0.8 million	12%

The internal rate of return can easily be calculated for these projects by dividing the equivalent annual net benefits by the investment, with the results shown in the final column of the table. Since the IRR is greater than the firm's MARR of 10% for projects B, C, and D, these projects are all acceptable, but Project A with its return of only 9% is unacceptable. If only one of these projects can be undertaken, then selection based upon IRR would choose project B – but didn't we just figure out that Project C was the best? What's going on? Why doesn't the IRR method result in the same choice as the NPV method? The fact that projects cannot simply be ranked by their IRR is a serious problem with using this measure.

A second problem with the IRR is that the process for estimating it could produce multiple answers. The problem of dueling IRRs could arise whenever the stream of annual cash flows switches from positive to negative more than once, which is why the IRR method is seldom praised by academics. In most projects presented to the board of directors, however, there will be a pretty clear initial investment that produces positive annual net benefits that continue for an indefinite period with at most a heavily discounted cost for decommissioning in the distant future. With such projects, there will be an unambiguous result, which is why this method is commonly used in business.

A third problem arises because the IRR methodology assumes that any cash received during the course of the project can be reinvested at the same IRR, while future costs can be discounted using the IRR. For projects with a very high IRR, both assumptions could be very unrealistic.

External Rate of Return

Since the IRR method is so commonly used in business, it is important to understand how to deal with these problems that might arise when it is used. A somewhat more complicated approach avoids the problem of multiple values for the internal rate of return as well as the difficulty of assuming that costs and benefits can be discounted with what could be a very high IRR. This approach uses what is called the **external rate of return** along with equivalence relationships to create an easily understandable comparison between costs and benefits:

- First, divide all of the periods considered in the analysis into periods where the cash flow is negative and periods where the cash flow is positive. There is no need to distinguish between investment costs, rehabilitation costs, or operating losses.

- Next consider the periods with negative cash flow. For each such period, we could establish a fund that would be expected to grow over time so that it could be used when needed to cover the negative cash flow. The size of the fund could be determined by using a discount rate that is consistent with the company's expected overall return on investment during the intervening years. This discount rate – the **external rate of return** – could perhaps be the company's minimum acceptable rate of return or the company's average rate of return. All of the negative cash flows could be converted to an equivalent present value using this external rate of return for a discount rate.

- Next, using the same external rate of return, all of the positive cash flows could be converted into a future value. The logic in extrapolating these funds to the future is that any extra cash generated by a specific project will be used to promote the overall activities of the company. For example, if the company has historically enjoyed a rate of return of about 10%, and if conditions in the future are expected to be no different, then the company could expect that the earnings from any new project in year t could be re-invested and earn 10% per year from year t until the end of the analysis period.

Figure 1 summarizes the process of calculating a project's ROI using the external rate of return method.

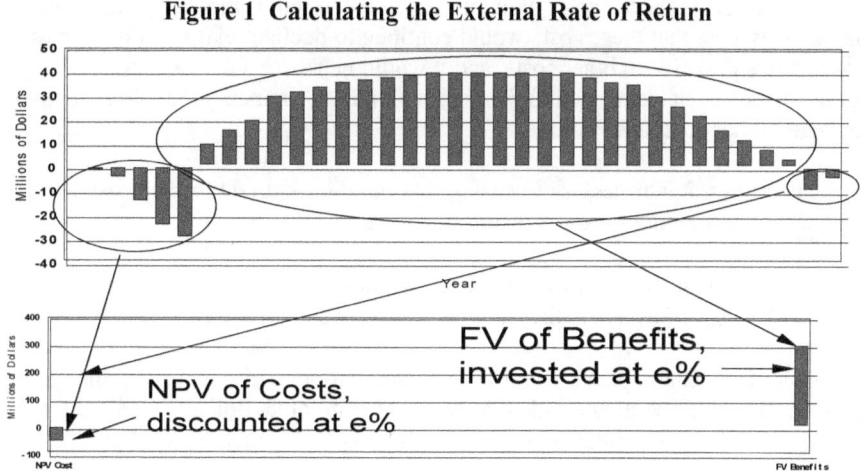

Figure 1 Calculating the External Rate of Return

The discount rate used in these calculations is referred to as the "external rate of return", where "external" indicates that the rate of return is based upon factors that are unrelated to the specific project that is being investigated. The same external rate of return would be used for evaluating any project; it is not something that would have to be defined for each specific project.

Given an external rate of return, the following comparison between the future value of the positive cash flows and the present value of the negative cash flows can then be used to determine the return on investment (ROI) for this project:

(Eq. 1)   FV positive cash flows = $(1 + ROI)^n$ (PV costs)

Before using this equation, the external rate of return "e" must be used to calculate both the present value of costs and the future value of benefits. The ROI that satisfies the equation can readily be obtained by trial and error using a spreadsheet. The ROI could by coincidence equal "e", but most likely will be higher or lower.[1]

---

[1] This section uses the term "external rate of return," which is the term commonly seen in engineering economic textbooks. ERR is presented as a measure that is similar to, and better than, the internal rate of return, but it is not clear what is really meant by "e", the external rate of return. However, consider a company that uses their own discount rate to determine the present worth of the

The external rate of return approach is favored by academics, as it avoids the necessity of implying unreasonable returns for reinvesting profits, and it provides a reasonable means of dealing with future periods with negative cash flow. However, this approach is unlikely to be encountered outside of textbooks. Public agencies are apt to consider ratios of benefits to cost rather than ROI, while private companies use the internal rate of return as an easier and apparently more objective result.

Constant Dollar vs. Current Dollar Analysis

The discount rate observed in the financial markets reflects three factors: the return available on risk free investments, a risk premium, and inflation. Any analysis that uses historical or projected interest rates is using data that reflects past and future expectations concerning inflation. Inflation expectations are also among the many factors that affect the price of stocks and real estate. Inflation is also a factor in MARRs of individuals, companies and government agencies.

It is important that inflation be treated consistently when evaluating projects. In estimating costs and revenues, it is often convenient to ignore inflation. So long as the major elements of cost and the major sources of revenue all increase at about the same rate, a constant dollar analysis will result in a reasonable projection of cash flows. If there are some components that are expected to perform much differently, then some adjustments would have to be made in projecting cash flows. For example, the costs of computers and communications have declined for decades, so that it would be reasonable to assume that these costs would continue to decline relative to other costs. Over the last 20 years, energy costs have risen relative to other costs, and it would be reasonable that this trend would continue. Any project that had significant costs related to communications, computers, or energy therefore might require adjustments in projections of constant dollar costs and revenues.

Given projections of cash flows, it is necessary to ensure that the discount rates are consistent with the assumptions about inflation. Two sets of assumptions are reasonable:

- Constant dollar: neither cash flows nor discount rates consider inflation
- Current dollar: cash flows and discount rates both reflect inflation

If cash flows are provided in constant dollars, but are discounted with real discount rates, then future cash flows will be discounted too much. If cash flows are provided in current (i.e. inflated) dollars, but are discounted with discount rates that do not consider inflation, then future cash flows will be insufficiently discounted.

Inflation in even the most stable economies during the most stable economic conditions will usually be at least 1-2% a year; in other circumstances, inflation easily could be 3-4% per year in the most stable economies and much higher elsewhere. This is not a factor to be overlooked, as the mistakes could be considerable.

Choosing Among Independent Investment Options

Consider a company that has many independent investment opportunities. These opportunities are independent in the sense that choosing any one of them does not require or preclude any of the others. The company could decide to choose none, any, or all of the options. In theory, the company could decide to invest in any project with positive NPV. If the NPV is positive, that means that the project will produce cash flows that will, when discounted at the company's MARR, be equivalent to having more money today. However, the company's MARR will be at least as great as its weighted average cost of capital, and the cost of capital conceivably could rise if the company attempted

---

costs of a project and the future worth of the benefits. "Return on investment" for a particular project would then be defined as the annual growth that would be needed for the present worth of project costs to grow into the future value of project benefits at the end of the project life. With this approach, there is no need to introduce "e" as something new, because the usual discount rate would be used in the calculation. The return on investment for the project would then be seen as a clearly defined measure that is naturally dependent upon the use of the proper discount rate, just as NPV is dependent upon the use of the proper discount rate.

to raise excessive amounts. The company's executives and board of directors would also have some concerns about the quality of the analysis and the possibility that some projects might prove to be less successful than they hoped for. As a result, the funds available for projects would likely be limited, and only the best projects would be chosen. The objective would therefore be to maximize NPV subject to a capital budget constraint, which would be equivalent to maximizing the return on investment for the capital that is budgeted.

In practice, many companies use the IRR in evaluating independent projects. The IRR can easily be used to rank all projects according to a common metric. The IRR will be deemed acceptable so long as it is higher than the companies MARR (usually viewed as the company's weighted average cost of capital). The decision process is straightforward (Figure 2), at least related to strictly financial matters: choose the projects with the highest IRR, so long as the IRR exceeds the hurdle rate and the total investment is within the budgeted amount.

Figure 2 Selecting Projects Based Upon a Hurdle Rate of Return

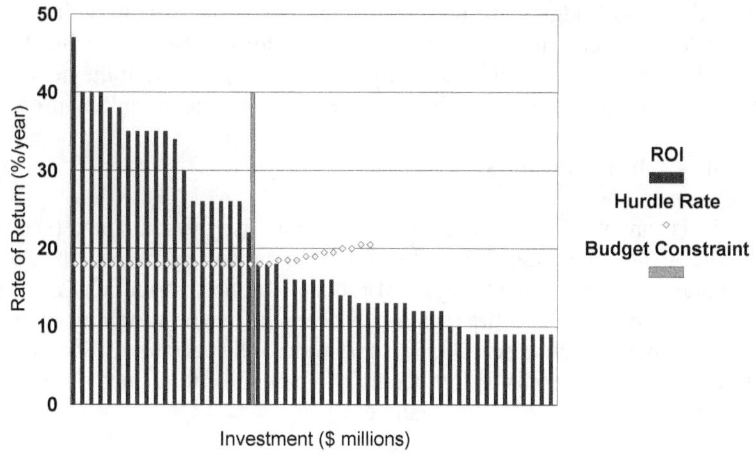

This process of ranking projects by their IRR assumes that the risks associated with the project are similar, so that they can each be compared to the same hurdle rate. A large company with many diverse opportunities for cutting costs or expanding markets will in fact have many investment options with similar risks: they know what to expect if they decide to make the investment. If the company is moving into a new type of business or if an investment is believed to have unusual risks, then a higher hurdle rate could be used. If an investment is deemed essential to the company's safety or to its continued operation, then the investment will be made even with an IRR lower than the hurdle rate.

As with so many elegant frameworks, this clear and logical process for selecting projects may not work so clearly or logically in practice. While a company indeed should know its MARR, that may be a subject of debate or it may not be something that is ever explicitly defined. The elegant model indicates that all projects whose expected return exceeds the MARR should be approved, ignoring the fact that there is always some kind of limit for capital expenditures. The limit is undoubtedly flexible, but that means that marginally acceptable projects may be approved only if they are supported by people with the power or persuasiveness necessary to convince the board of directors. The decision model depicted above also assumes that there is an ordered list of all of the feasible projects, none of which are mutually exclusive. No one who has ever seriously considered design will assume that they can ever know all of the alternatives, many of which will certainly be mutually exclusive. In a large organization, whether public or private, leaders from each department will be promoting their own projects; those who are more diligent, more eloquent, or closer to the senior officials may be the ones whose projects are approved.

There may well be many better projects that no one thought of or that no one wanted to champion. If you are an analyst or a consultant or a reviewer of a project, it is your job to look for some of those other options. Some possibilities would include:

- Use of better materials and techniques to build the same facility
- A better structural design to serve the same purpose
- A different location for a similar project
- A different scale – many smaller projects or fewer larger projects

In general, no one can prove that their design or proposed project is the best. They can only defend, refine, or abandon their proposal in response to whatever feedback and opposition they receive.

Ranking Independent Projects using Present Value, Future Value, Annual Value or IRR

If making money is your objective, then ranking projects by present, future or annual value would certainly seem to be the correct approach. The rankings obtained by using any of these three approaches would be the same, as the differences among them depend upon factors that vary only with the discount rate and the period of the investments. If the projects are independent and budgets are unlimited, then any project with positive NPV would be worth pursuing. If the NPV is greater than zero, then the internal rate of return will be greater than the discount rate, so that the IRR will also identify which projects should be pursued. However, as shown in the next subsection, the projects will not always be independent, and that is when difficulties are likely to arise in ranking projects using the IRR.

Choosing Among Mutually Exclusive Projects

Sometimes competing projects cannot all be pursued. They may both use the same land (should we build a hotel or an office building on this site?), they may offer different solutions to the same problem (should we build a bridge or a tunnel?), or they may be related to competing strategies of production and distribution (small retail outlets in every neighborhood or large box stores to serve the entire region?). There will also be variations in projects related to design and scale of effort: should the sports stadium seat 30, 50, or 75 thousand spectators? Should the bridge have four or six lanes? Should apartments have four rooms or five rooms? In cases like this, once a particular design is selected, the others are no longer available; the choices are mutually exclusive.

When selecting from a group of mutually exclusive projects, it does make sense (from a financial perspective) to maximize the net present value of cash flow. The best project will indeed be the one that is equivalent to the largest amount of money today.

However, if a company evaluates projects by choosing the ones with the highest IRR, problems are likely to arise. A prior example has already shown that the project with the highest IRR may not be the best project. The following example shows that the key to properly using IRR is to consider the rate of return on each increment of investment. If an incremental investment exceeds the MARR, then that increment can be justified, even if it lowers the IRR for the project.

Table 2 summarizes the investment and expected annual net income for four options for developing a site: build a parking lot or construct a building with one, two or three stories. If we assume that the same net income would continue indefinitely, then the annual rate of return would be the net income divided by the investment. For example, the parking lot's rate of return would be $22,000/$200,000 = 11\%$. We could also estimate the present worth of the income using the capital worth method: present worth equals annual income divided by our discount rate, and the NPV of the project would be the present worth of the income minus the investment cost. For the parking lot, assuming a 10% discount rate, the NPV would be $22,000/10\% - \$200,000 = \$20,000$. Table 3 shows the rate of return and the present worth for these four options.

Table 2 Mutually Exclusive Options for Developing a Site

Project	Investment Required	Annual Net Income
Parking lot	$200,000	$22,000
One-story building	$4,000,000	$600,000
Two-story building	$5,500,000	$720,000
Three-story building	$7,500,000	$960,000

Table 3 IRR and NPV for the Mutually Exclusive Options for Developing a Site

Project	Internal Rate of Return	Net Present Value
Parking lot	11.0%	$20
One-story building	15.0%	$200
Two-story building	13.1%	$170
Three-story building	12.8%	$210

If our hurdle rate equals our discount rate of 10%, then all of the projects are acceptable, whether we consider the IRR or the Net Present Value. However, if we have to choose just one of these, then we have a problem. Considering only the return on investment, the best choice appears to be the one-story building with its 15% return. However, the three-story building has a higher net present value. Which is really the best project? In your presentation to the board of directors, do you recommend the one-story building because the company always uses IRR to rank projects? Do you recommend the three-story building because the text books always recommend maximizing NPV? Do you accept a suggestion to compromise on a two-story building? What should you do?

To deal with these questions, it is necessary to do the analysis one step at a time, beginning with the option that requires the least investment, which in this case is the parking lot. Since the IRR of this project exceeds the hurdle rate of 10%, it is acceptable. The question now concerns the additional benefits that might be obtained from additional investment in this site. The one-story building requires an additional $3.8 million dollar investment in order to gain an additional $578 thousand in annual income. The rate of return for this incremental investment is therefore $578/$3800 = 15.2%, which is well above the hurdle rate. The NPV for this building is ten times greater than the NPV for the parking lot, so both measures indicate that the one-story building would be a good investment. Now we need to consider the benefit to be gained by the additional investment required to go from one to two stories. The additional investment of $1.5 million produces additional net income of $120,000 per year, so the ROI for the increment is only 8%, which is less than our hurdle rate of 10%. Thus, the two-story building is not as good as the one-story building. Although the IRR for this building is 13.1%, which is well above the hurdle rate, the incremental return for the additional $1.5 million is unacceptable. If we just look at the NPV, we immediately reject the two-story building because the NPV is $30,000 less than the NPV of the one-story building.

Now we proceed to the fourth and final option, the three-story building. We compare this building to the best of the previous options, namely the one-story building. The incremental investment in this case is $3.5 million and the incremental net income is $360 thousand, so the incremental return is 10.3%, which is just over the 10% hurdle rate. Therefore, the incremental investment is in fact justifiable. Once again, the NPV immediately gives the same result: the three-story building has the highest NPV and therefore is the preferred investment.

It is conceivable that the board of directors might be unwilling to commit $7.5 million to this site. If so, someone who hadn't followed the logic very closely might suggest cutting back to the two-story building, which after all has an IRR of 13.1% (perhaps snidely noting that this is higher than the 12.8% for the option recommended by the junior analyst). That would be the point in the meeting where you have to stand your ground: if the board is unwilling to commit $7.5 million, then they should stay with the one-story building because it has a higher NPV than the two-story building. And, if necessary, explain that the incremental $1.5 investment required for the two-story building would be better invested in another of the company's projects.

The procedure illustrated in this example can be used with any set of mutually exclusive investment alternatives:

- Rank the alternatives in increasing order of investment required.
- Estimate the IRR for the each alternative.
- Choose as a base case the first alternative whose IRR exceeds the hurdle rate.
- Compare the next alternative (i.e. the alternative with the next highest investment requirement) to the base case:
    - Calculate the IRR for the incremental investment.
    - If the incremental IRR is unacceptable, consider the next alternative and repeat this step.
    - If the incremental IRR is acceptable, make this alternative the new base case and repeat this step until either the capital budget is reached or all alternatives have been tested.

This process will find the highest investment that can be justified among the competing projects. It will also select the project with the highest NPV – which is why it is desirable to estimate the NPV even when you must present results in terms of IRR.

Dealing with Unequal Lives of Competing Projects

Competing projects may well have different project lives. If so, then several approaches can be taken to ensure that comparisons are done in a reasonable manner.

One possibility would be to choose a longer period that is an integral multiple of the lives expected for each of the projects. It could then be assumed that the projects would be repeated two or more times over the course of extended period of analysis. For example, if competing projects have lives of three years or four years, then the analysis of each could be done for a period of 12 years, as this would involve four cycles of the three-year projects and three cycles of the four-year projects. This approach could lead to some extraordinarily long life cycles if there are many projects with many different lives. For example, if projects have lives of five, seven or ten years, then a 70-year project life would be needed to have an integral number of cycles for each project. The problem with such a long project life is that it is very likely that technology, population, related development, and prices would change so much as to make very long-term estimates very questionable. It is not reasonable to use a 70-year horizon to compare options that all have lives of at most ten years.

A second approach would be to use the annuity value rather than the net present value. The assumption underlying this approach would be that any of the projects could either be extended at the same or a similar annuity value or be replaced by better projects. If the projects have similar, but not identical lives, the differences will not be dramatic:

> *Fundamentally, Equivalent Annual Cost is a robust measure regardless of the alterations from the original project and its identical repetition assumption. ... In reality, projects often do not repeat, but are rarely divested during their first life and dramatic cost change occurs only in the long run.*[2]

For typical projects, where the effect of unexpected early termination is minor and discount rates exceed 10%, the equivalent annual cost is reasonable to use even though projects have different lives. For riskier projects or projects with great uncertainty in cash flows, sensitivity analysis must be done to consider the effect of early termination and variable cash flows on the equivalent annual cost.[3]

---

[2] Ted G. Eschenbach, Robert B. Koplar, and Alice E. Smith, *"Violating the Identical Repetition Assumption of EAC"*, **1990 International Industrial Engineering Conference Proceedings,** Institute of Industrial Engineers, pp. 99-103

[3] Idem.

A third approach would be to include a residual value for each project at the end of the analysis period. The assumption underlying this approach is that it is possible to estimate residual values, which may be feasible, but which may also be much more trouble that it is worth.

A fourth approach is simply to use a long enough time period that any differences would be minimal. If discount rates are greater than 10%, then what happens after 20-30 years will have minimal impact on NPV.

As always, it is important to use common sense. When in doubt, do some sensitivity analysis using different time periods to determine to what extent, if any, the choice of the period of analysis is causing differences in rankings among the alternatives. There is no "right" method that must be followed.

Splitting a Project into Pieces for Different Parties

So far, we have considered the perspective of entrepreneurs, developers, companies, or agencies as they evaluate their options for undertaking construction projects. The discount rates and hurdle rates that they use will reflect their own investment opportunities, their own cost of capital, and their own perceptions of the risks associated with the projects that they are examining. If their projects are funded entirely by cash on hand, then this is the only perspective that matters.

More commonly, financing a project is only possible if a major portion of the money required for the investment can be raised from outside investors. If this money is a small portion of the total funds sought by the company, then the cost of the capital required (i.e. the interest rate on loans or bonds and the price per share of stock that is sold) can be assumed to relate to the overall financial strength of the company or agency. The discount rate used in the calculation of the present worth and the hurdle rate would be at least as high as the organization's weighted average cost of capital, and the financing of any particular project would be a small part of the overall financial management of the company or the agency.

Additional analysis will be necessary if the project is undertaken as a stand-along activity of a new company, if the project requires funding that is tied to its actual results (rather than to the overall financial strength or the organization), or if the project is a major departure from prior activities of the company. The project may require loans from a bank that are secured by the expected rents, tolls, or other proceeds of the project. The value of the company's stock could be related directly to the success of the project, taking into account the interest payments which must be paid to banks or bondholders before any dividends can be paid to stockholders. In these situations, it is necessary to consider the different perspectives of the potential investors.

A mortgage is a loan that is secured by a lien on the property. If mortgage payments are not made in a timely fashion, the mortgage holder has the right to foreclose on the property. Since the loan is backed by property, the mortgage is less risky than an unsecured loan, and the interest payments on a mortgage will be lower than the interest on an unsecured loan. It is possible to have multiple mortgages on a property. If so, then the mortgage agreements will state the order in which payments will be made if there is insufficient cash to make all of the contractual payments. The first mortgage will generally have priority over the second or any other mortgages, meaning that the holder of the first mortgage has first call on the cash flows of the company. The risk of not getting paid is therefore higher for the holder of a second mortgage than for the holder of the first mortgage.

After the payments are made on secured loans, the next priority will be to make payments on unsecured loans and to pay interest on bonds. If a company is unable to make such payments, then it can be forced to declare bankruptcy. A bankrupt company can in many cases suspend mortgage payments, interest payments, taxes and other fixed charges in order to reduce the outflow of cash while attempting to reorganize.

After all of the fixed charges and taxes are covered, whatever cash is left over can be paid out as dividends to stockholders or re-invested in the company. This portion of the cash flow will vary with the success of the company or the project; the higher the fixed charges as a total proportion of expected cash flows, the more uncertain the prospects for the company. The value of the company to the owners depends upon this portion of the cash flow: the higher and the more reliable the cash flow, the greater the value of the company.

Summary

Maximize the Net Present Value of Cash Flows

The equivalent worth methods provide the best way to compare alternatives. If the net present value is positive, then a project is worth pursuing, at least from the financial perspective. If the net present value is negative, then it is not worth pursuing. If the net present value is positive, then any future values and annuity values will also be positive, so any of these measures can be used to determine whether or not a project is worthwhile from the financial or economic perspective. Moreover, each of these measures will produce the same ranking of independent alternatives and the same choice among mutually exclusive alternatives.

Using the Internal Rate of Return to Rank Projects

Companies commonly use a different measure, the internal rate of return, to rank competing proposals for projects. The internal rate of return is useful because it can be calculated without reference to any pre-determined discount rate. It therefore appears to provide an objective assessment and an obvious means of ranking independent alternatives. However, there are three potential problems in using this measure to rank projects:

- If cash flows are highly variable, with multiple periods where cash flows are negative, then the methods used to estimate IRR may come up with two values.
- This method implies that all positive cash flows can be reinvested at the IRR over the life of the project. In fact, cash obtained from the project may have to be invested in ways that have much different returns.
- This method does not provide the correct rankings for mutually exclusive alternatives. It is necessary to consider the incremental return for incremental investments to determine which of such alternatives is best.

The IRR can be modified to deal with the first two problems by using an external rate of return for converting all periods with negative cash flow to an equivalent present value and converting all periods with positive cash flow to an equivalent future value. The rate of return is then uniquely defined as the annual return that will cause the present value of the costs to grow into the future value of the benefits. This approach will still require analysis of incremental returns for incremental investments in order to obtain the correct selection among mutually exclusive alternatives.

The Importance of Project Life

When comparing alternatives, the time period needs to be chosen with some care. In general, a period of 20-30 years will be sufficient because the discounted costs and benefits of more distant benefits will add very little to the present worth of a project. The choice of a time period should not determine the outcome of the analysis. If costs and benefits have both reached a steady state by the end of the analysis period, then there is no reason to worry about the choice of the project life, so long as the discounted cash flows from the excluded years contribute little or nothing to the present worth. If either costs or benefits are expected to rise or fall sharply just after the end of the analysis period, then a longer period will be needed. For example, if extensive rehabilitation is anticipated around 22-25 years, then the use of a 20-year life could be very misleading and a 30-year life would be better. In some cases, where there are extraordinary costs or benefits in the far distant future, much longer time periods should be considered. For example, the costs of dismantling an obsolete nuclear power plant and the ultimate costs of safely disposing or sequestering spent nuclear fuel should be included in the analysis, even if such costs are expected only after 40 or more years of operations.

Choosing Among Mutually Exclusive Alternatives.

If selecting one option precludes other options, then the options are mutually exclusive. The basic rule is to choose the option with the highest NPV (which will also have the highest annuity value and the highest future value),

If the internal rate of return is used, then care must be taken in ranking projects, because the project with the highest IRR will not necessarily be the best project. Constructing a smaller project with a higher IRR may preclude a larger project with a lower, but still acceptable IRR. It is therefore necessary to follow a well-defined procedure in determining which project is best.

- Rank the mutually exclusive projects in order of increasing investment requirements
- Determine the IRR for each project
- Starting with the smallest project, select the first project with an acceptable IRR as the base project
- Calculate the incremental costs and incremental benefits for each larger project.
- Calculate the incremental rate of return for each larger project.
- Select the first project larger than the base project that has an acceptable incremental rate of return. This becomes the new base project.
- Repeat the analysis of incremental costs relative the most recent base project.
- If there are no projects for which the incremental benefits justify the incremental costs, then the most recent base project is the best project.

Discussion: the Limits of Financial Analysis

In the private sector, financial performance will usually be what is most important in project evaluation. However, in the public sector, where projects are undertaken to meet public needs rather than to make a profit, economic, social, environmental, and sustainability issues will also be relevant. To the extent that economic factors can be expressed in monetary terms, it will be possible to use the same methodologies to calculate the NPV and the IRR. However, these global measures will not be the only things to consider when evaluating any complex project. Other economic issues will include distributional equity (who wins and who loses), regional economic impact (the use of local labor and resources and the multiplier effect on the local economy), and non-financial externalities (environmental and social impacts and the need for remediation). Any large project will have an impact on the public, and there will likely be many costs and benefits that are difficult to quantify and even more difficult to value. In some cases, non-quantifiable factors will be the major issues in project evaluation.

In conclusion, despite spending a great deal of time focusing on financial matters and believing that financial feasibility is essential for any project, we must recognize that financial feasibility may have little or nothing to do with project desirability. Whether or not it is possible to get money to build something is much different from whether or not something should be built. Financing difficulties may preclude certain highly desirable projects, yet encourage other clearly undesirable projects.

Engineers, managers, planners, and politicians have some personal responsibility for pursuing desirable projects that are financially feasible. Project evaluation depends upon proper presentation of estimated costs and benefits and disclosure of assumptions concerning discount rates, project lives, and the types and distribution of costs and benefits. It is not enough to show that a particular project can be justified; it is also necessary to show that it is better than the available alternatives.

Rules of the Game: Taxes, Depreciation and Regulation

The zoning law of 1916 – the nation's first – regulated the bulk of buildings, their height, and their uses. It divided the city into three zones – residential, business, and unrestricted – and empowered the Board of Estimate to regulate the use, height, and bulk of every building on every street in the city, depending upon what zone the block was in.[1]

Introduction

Government policies affect how projects are conceived, what kinds of projects can or cannot be implemented, where projects can or cannot be built, how projects can or cannot be constructed, and how successful they will be once they are implemented. The most important policies relate to taxes, land use, safety, and the environment. These policies are in effect "rules of the game" that limit what kinds of developments can be pursued and influence how the players tabulate their scores, i.e. their profits and their ability to complete projects. The players – real estate companies, entrepreneurs, public agencies, infrastructure operators, investors, banks – still have to figure out what they want to do and how to do it, but they must abide by the rules that have been established. Changing the rules will change the way that the game is played, and the rules can be adjusted to promote projects that are believed to provide economic, social, environmental or sustainability benefits for the public.

Taxes are relevant to project evaluation because taxes affect cash flows. If the goal is to maximize the net present value of cash flows, then it is necessary to consider taxes. Moreover, local, state and federal governments may impose taxes or offer tax credits in order to discourage or promote certain kinds of development, so it is important to be able to comprehend the effects of tax policy on project design and evaluation. Different kinds of taxes may apply to the profits from constructing, operating, and selling infrastructure, so it is necessary to understand how tax laws categorize each type of expenditure and each type of revenue. Accounting rules established by law or by regulation determine what kinds of expenses are treated as current expenses and what kinds are treated as capital expenditures. Arcane rules may determine whether money spent on rehabilitating infrastructure is treated as operating expense – which is fully deductible as an expense in the current year – or a capital expense that can only be deducted from taxable income over a period of many years.

Zoning is the major tool used by local governments to guide land use. Zoning restrictions limit the types of development that may be pursued in certain locations, such as restricting one area to residential use while specifying another area as suitable for industrial use. Zoning restrictions may also limit the size or height of buildings or the location of buildings on a site.

Building codes define what types of construction materials, designs, and methods can be utilized. Regulations can be established to reduce risks during construction or operation or decommissioning of a project. New technologies, such as the use of plastic pipes for plumbing, had to be approved for use in building codes before they could be widely used. In transportation, governments may create design standards for highways and bridges that serve a similar function, i.e. promotion of safety during construction and operation. Likewise, governments may establish standards for the construction and operation of water resource systems and for other types of infrastructure.

Environmental restrictions restrict the nature, location and cost of projects. Land use regulations may include restrictions on development in or near wetlands or waterways. There may be restrictions on the types of materials that are used in construction, such as laws that prohibit the use of asbestos because of the link between asbestos and lung cancer. There may also be restrictions that limit the types of work that can be undertaken at night, so as to limit the

---

[1] New York City implemented zoning to avoid turning Manhattan into dark canyons, with skyscrapers towering above and keeping light and fresh air away from city streets. John Tauranac, **The Empire State Building: The Making of a Landmark**, St. Martin's Griffin, NY, NY, 1995, p. 55

disruption to neighborhoods. Contractors may be required to take special precautions to prevent dust and run-off from construction sites from contaminating nearby areas.

Depreciation and Taxes

Taxes and financial statements are structured according to strict guidelines known as generally accepted accounting principles. By following these guidelines, it is possible to define terms such as profit and return on investment (ROI). Profit, ROI and other measures used in financial statements are critical because investors and analysts use these measures in judging whether or not to invest in a company. The financial markets rely upon the validity and the comparability of data produced by companies in their financial statements. Therefore, if a company sells stocks or bonds, it is required to use accepted accounting principles in preparing those statements, as well as in preparing their tax returns.

Since taxes are large cash flows, they cannot be ignored when evaluating projects. And, since the amount of taxes to be paid depends upon accounting rules, it is necessary to understand some basic concepts of accounting. One of the most important rules is that capital expenditures cannot be treated as a current expense, but instead must be spread out over many years as a depreciation expense.

For example, operating expenses for a building include such things as electricity for lights, oil for heat and wages for the people who manage and maintain the building. At the end of each week or month, the owner of the building knows how much was spent on electricity, oil and wages: the lights worked, the building was warm, the rooms and hallways were cleaned, and rents were collected. The money was spent, the work was done, and now it's time to do it again in the next month.

The capital expenditures that were required to create the building are entirely different. The building may have cost $10 million to construct, and when the construction was complete the owner may have a mortgage for $10 million – but the owner also has a building. Now, the building may or may not be worth $10 million, because the value of the building depends upon the real estate market, the condition of the building, and the annual rent payments, not the cost of the building. However, the accounting assumption is that if the building cost $10 million to construct, then the building is an asset worth $10 million when it is put into service. The owners may have spent $10 million, but they have created an asset worth $10 million, so they have not had any loss in value.

The same concept could be applied to the purchase of a car for $20,000, the construction of a bus terminal for $20 million, or the construction of a vast pipeline for $2 billion. The money may have been spent, but an asset has been created, and accountants will record the book value of the asset as being equal to the investment cost. So right at the beginning of the life of the asset we have an accounting assumption that is accepted, even though it most likely is wrong. The car may be worth only $15,000 as soon as we drive it out of the dealer's lot; the bus terminal may only be worth $10 million to anyone other than the bus company, while it may be worth $30 million to the bus company. We don't quibble about this discrepancy between book value and real value, not because we are so flexible in our thinking, but because the tax collector tells us that we will use the accountant's book value as one factor in computing the taxes we owe. If you want to use the real value of the asset in your own internal reporting, that is fine; just don't confuse the book value and the real value when doing your taxes. Also, although we really do know what we paid for the car, the bus terminal and the pipeline, we probably do not know what they really are worth today. So, it will be convenient for us, as well as the tax collector, to use the accounting assumption.

Now we have a place to start for figuring out depreciation, namely the book value of the asset. We could perhaps now try to determine how much the asset deteriorates each year, so we could use the actual decline in life as the amount of depreciation. This turns out to be a difficult task. What is the expected life of a car? How much of the life of a car is consumed by time? By usage? By exposure to rain and snow? It is possible to do some engineering analyses to answer questions like these, but the tangle of assumptions and analyses will quickly become quite thick. For internal purposes, say for a rental car company, it might be a very good thing to understand the life of cars based upon the kind of usage they receive and whether they are based in the heat of Florida or the snow and ice of Minnesota. For most companies, however, it would probably be a difficult calculation with little or no benefit to management; a plan to replace company cars after five years will be sufficient for them. However, all companies will have to use an accepted methodology to account for depreciation of their cars.

At this point, we need three more accounting assumptions to make it easy to estimate depreciation:

- The life of the asset
- The salvage value of the asset (which would not include the value of land, as land is generally assumed not to depreciate in value)
- The depreciation rate over the life of the asset

The life of the asset could be defined based upon some sort of study of past experience – or it could be defined by the tax collector. A car, for example, might be assumed to have a life of ten years, while a bus terminal or a pipeline might be assumed to have a life of thirty or fifty years. The salvage value is the remaining value of the asset at the end of its life. For a car, the salvage value might be the scrap value of the car, which might be assumed to be 5% of the original purchase price. For a bus terminal or a pipeline, the salvage value might be assumed to be the book value of the land (i.e. the purchase price of the land required for the project) or a percentage of the investment cost. The accounting principle is that depreciation causes the book value of an asset to decline from the original investment cost to its salvage value at the end of the asset's life. The depreciation could most easily be assumed to proceed at a constant rate over the life of the asset:

(Eq. 1)   Annual Depreciation = (Investment – Salvage)/Life

If the life of a $4 million asset is ten years and the salvage value is $1 million, then the asset would depreciate in value from $4 million to $1 million over its ten-year life, for an average depreciation rate of $0.3 million per year (Figure 1). This is called "**straight-line depreciation**", because the book value of the asset declines in a straight line from the initial investment cost to the salvage value.

Figure 1 Straight-Line Depreciation

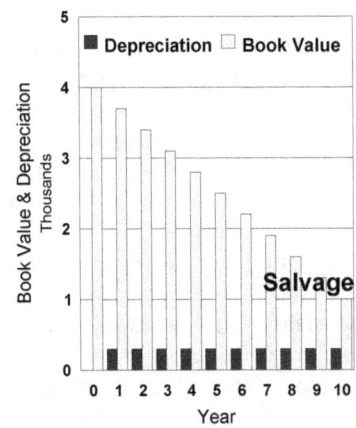

Other approaches to depreciation can be imagined. One might argue that depreciation should be greater at first, to reflect the immediate loss in value of cars and some other assets. Once we have accepted that the accounting assumptions may be tied to convenience rather than reality, it is easy to come up with some alternatives. For example, instead of assuming a fixed amount of depreciation over the life of the asset, it would be possible each year to take a fixed percentage (P%) of the remaining book value over the life of the asset, without considering the salvage value at all.

(Eq. 2) First year depreciation = (P%)(Investment)

(Eq. 3) Reduced book value at end of first year = (1-P%) (Investment)

(Eq. 4) Second year depreciation = (P%)(1-P%)(Investment)

(Eq. 5) Year N depreciation = P% (Book value at end of year N-1)

These equations can readily be used to create a table in a spreadsheet that shows the initial book value in each year, the amount of depreciation allowed in that year, and the final book value.

With the same example used above, suppose the book value was depreciated by twice the amount allowed under straight line depreciation, which would be 20% of the remaining value each year for ten years. This method is called **double declining balance depreciation**, because the amount of depreciation allowed in the first year is double the amount allowed under straight line depreciation. The book value would decline by (0.20) $4 million = $800,000 in the first year. In the second year, the initial book value would therefore be $3.2 million, and it would decline by (0.20) $3.2 million = $640,000 in the second year. Each subsequent year the amount of the depreciation would decline by a smaller amount, as illustrated in Figure 2. For the first five years, the annual depreciation is higher using the double declining balance, but after that the fixed depreciation of $0.3 million under the straight line approach is better.

This method can be used with other percentages, and the general method is known as **declining balance depreciation**.

Figure 2 Double Declining Balance Depreciation

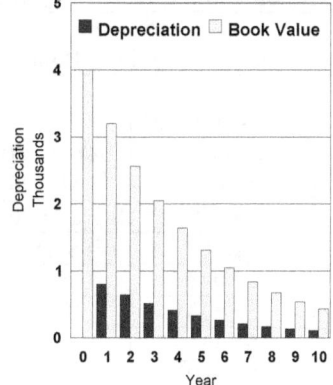

A related approach is to use the declining balance approach only so long as that approach results in depreciation greater than the straight-line approach. Using this method, it is necessary to compare the depreciation for the declining balance approach with the depreciation that would be obtained by switching to straight-line depreciation of the remaining book value over the remaining life. Once the straight line approach provides greater depreciation, switch to that approach and continue over the life of the project. Once again, this method can readily be examined in a spreadsheet:

- Start by using the declining balance in the first year.
- For the second year, calculate depreciation using two methods:
    o Continue using the declining balance method.
    o Determine the depreciation of the book value at the beginning of the year using the straight line method (Eq. 1).
- If the depreciation under the straight line method is higher than that allowed under the declining balance method, then switch to that method; if not, then continue to use the declining balance method.

This method is illustrated in Figure 3. Table 1 compares the book value and the annual depreciation for the three methods. Like the double declining balance method, this approach provides a way to accelerate depreciation of an asset, thereby allowing larger tax deductions during the early years of an assets life.

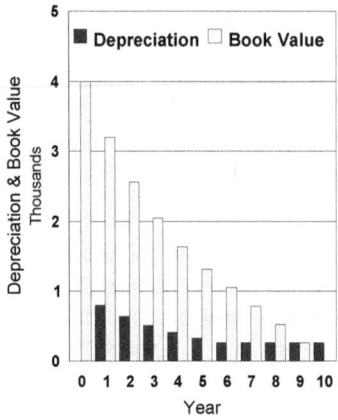

Figure 3 Double Declining Balance Reverting to Straight Line Depreciation in Year 6

Table 1 Three Methods of Depreciating an Asset With Initial Book Value of $4 Million and a Life of Ten Years ($ millions)

Year	Straight Line		Double Declining Balance		Double Declining Balance Reverting to Straight Line	
	Book Value at Beginning of Year	Depreciation	Book Value at Beginning of Year	Depreciation	Book Value at Beginning of Year	Depreciation
1	$4.000	$0.300	$4.000	$0.800	$4.000	$0.800
2	$3.700	$0.300	$3.200	$0.640	$3.200	$0.640
3	$3.400	$0.300	$2.560	$0.512	$2.560	$0.512
4	$3.100	$0.300	$2.048	$0.410	$2.048	$0.410
5	$2.800	$0.300	$1.638	$0.328	$1.638	$0.328
6	$2.500	$0.300	$1.311	$0.262	$1.311	$0.2621
7	$2.200	$0.300	$1.049	$0.210	$1.049	$0.2621
8	$1.900	$0.300	$0.839	$0.168	$0.786	$0.2621
9	$1.600	$0.300	$0.671	$0.134	$0.524	$0.2621
10	$1.300	$0.300	$0.537	$0.107	$0.262	$0.2621
11	$1.000		$0.429		$0.000	

One might think that these rather arbitrary methods would require a great deal of bookkeeping, plus considerable trouble figuring out what the life of an asset is. That is true. And the accountants, legislators, and tax collectors in the U.S. have therefore come up with an even simpler approach for accelerated depreciation of assets. The U.S. Internal Revenue Service allows companies to use what is called the "Modified Accelerated Cost Recovery System (MACRS)," which was introduced in 1986. This system divides assets into six categories, defines an asset life for each category, and assumes the salvage value is zero. Another simplifying assumption is that the first and last years of an assets life are assumed to be exactly six months, so it is not necessary to track the actual date that assets were put into service. Since there are only a limited number of options, standard lengths for the first and last years, no salvage value, and no need to determine lives, book-keeping is simplified and there is no need to justify the choice of an asset life. As noted above, the term "accelerated" means that the lives in the MACRS are generally shorter than what were previously used, so that companies can take more depreciation sooner. The option for using straight line depreciation is still available for some assets.

Income Taxes

Individuals and companies pay income taxes that are calculated as a percentage of annual income. Income taxes may be progressive, which means that the amount of tax increases as a percentage of income for higher levels of income. The maximum tax rates vary from country to country, and a country may raise or lower tax rates from time to time. In the United States, additional income taxes are imposed by most of the states and some of the largest cities.

State and local taxes are deductible expenses when calculating federal income taxes, so that the effective income tax rate will be:

(Eq. 6)   Effective Income Tax Rate = SR + LR + FR(1 – SR - LR)

Where:

SR = state income tax rate

LR = local income tax rate

FR = federal income rate

Consider a company that pays federal taxes at a rate of 34% plus state taxes of 6% and a city tax of 1%. Their effective income tax rate will be:

(Eq. 7)   Effective Income Tax Rate = 6% + 1% + 34%(1-.06-.01) = 39.55%

Income taxes can have a large impact of cash flows, and it is necessary to consider how taxes relate to profits, return on investment, internal rate of return (IRR) or the minimum acceptable rate of return (MARR) for a company or an individual. Such financial objectives can be calculated before or after income taxes. The relationship between the before-tax and after-tax MARR and IRR can be approximated as follows:

(Eq. 8)   After-tax MARR = (1- Effective Income Tax Rate) (MARR)

(Eq. 9)   After tax IRR = (1 – Effective Income Tax Rate) (IRR)

The higher one's tax rate, the more important tax effects become. Tax effects are extremely important for certain kinds of investment decisions. For example, in the United States, tax laws allow state and local governments to sell bonds whose interest is not subject to federal tax. Investors who live in the state that issues such bonds will not have to pay any state tax on the interest received from such bonds. Suppose a group of investors are considering municipal bonds instead of buying an AA-rated corporate bond that pays 6% interest. Would these investors be willing to buy a municipal bond offered by their local port authority that pays 4.0% interest? The investors agree that both bonds are very high quality, as they are rated double-A by bond-rating services. Some of the investors are in the 35% tax bracket, while others are in the 15% tax bracket; all of them pay 6% for state income taxes.

First consider the high income investors. Their effective income tax rate is calculated as follows:

(Eq. 10)   Effective tax rate = 6% + 35%(1-.06) = 38.9%

For these investors, the 6% interest received from the corporate bond will be taxed at the effective rate of 38.9%, so that they will only receive 6%*(1-0.389) = 3.67% interest after deducting federal and state taxes. They therefore would prefer the tax-free interest of 4% that they could obtain from buying the local bonds offered by the port authority.

For the lower income investors, the corporate bonds provide a better option, as these investors have a much lower effective tax rate:

(Eq. 11)   Effective tax rate = 6% + 15% (1-.06) = 20.1%

Their after-tax yield on the corporate bonds would therefore be 6% (1-.201) = 4.79%, which is well above the tax-free yield on the port authority's bond.

This example illustrates how the tax code can encourage certain investors to buy bonds to finance what legislators believe to be desirable projects. If the interest on municipal bonds were not deductible, then municipalities and states would have to offer much higher interest rates to attract investors.

Another way to promote projects is to allow companies and individuals to reduce their income taxes by an amount equal to a specified percentage of qualified investments. Governments may offer **tax credits** in order to promote investment in certain kinds of activities, such as education, housing for the elderly, rail transportation, or alternative energy programs. Legislation defines the type of expenditures that would qualify for the program and the amount of the tax credit. Infrastructure programs can be promoted via an **investment tax credit (ITC)** that could be used to reduce taxes during the year of the investment. Since infrastructure will be depreciated over a life of many years, the investment tax credit can be a major boon to developers during the first year of their projects.

For example, suppose the federal government has decided to provide a 20% investment tax credit to promote investment in housing for low income families. If a developer were to construct such housing at a cost of $4 million, then the tax credit would be $800,000. If the investment occurred in year 1, and if the company adjusted its estimated tax payments to reflect the tax credit, then the benefit would also be received during year 1, reducing the cost of the project from $4 million to approximately $3.2 million. Of course, the tax credit is a benefit only if the company actually pays taxes that are at least as great as the credit. If the company only paid $200,000 per year in federal taxes, then the tax credit would have to be taken over a period of 4 years. In this case, the tax credit would be very much like an annuity of $200,000 per year for four years, and it would be worth considerably less than $800,000. Thus, the tax credit will be most valuable to profitable companies that will immediately be able to use the credit to reduce their current taxes.

This example and the prior examples have illustrated how accounting rules and the tax code define what expenses can be deducted from revenue in order to calculate income and what portion of income must be paid as income tax. In particular, we have seen that rules related to depreciation can have a major impact on what will be considered to be profit and what companies have to pay in taxes. Since depreciation is such an important policy tool, it is worth considering in some detail how depreciation and tax policy work together to affect projects. There are four things to consider:

- Depreciation converts investments into recurring expenses. The investment involves actual cash expenditures (possibly using borrowed money) that are not deductible for income tax purposes, while depreciation is a non-cash expense that is deductible.
- Accelerated depreciation allows a business to increase depreciation during the early years of an asset's life. This accounting change has the effect of reducing profits, but increasing cash flow.
- Depreciation can be treated as an expense even if it is clear that the asset is actually increasing in value, as is often the case with buildings.
- If an asset is sold, the new owner can depreciate the asset again, using the purchase price as the initial book value and using a life and salvage value allowed by accounting rules and tax regulations.

Table 2 shows the tax advantages of the three options for depreciating the asset that was included in Table 1 above. It is assumed that the business's effective tax rate is 40% and that the business indeed has taxable income remaining whichever method is used. For straight line depreciation, the annual tax benefits equal $120,000, which is 40% of the annual depreciation of $300,000. Over ten years, the tax benefit is $1.2 million, and the net present value of the tax benefit, discounted at 10%, is $0.737 million. With the double declining balance method, the annual tax benefits are initially much higher, but decline over the ten-year life of the asset. The total tax benefit is greater, with a NPV of $1.023 million. The third method provides even more tax benefits, as this method fully depreciates the asset (i.e. no salvage value) and the NPV of the tax benefit is $1.096 million. The accelerated methods of computing depreciation thus could increase the value of this asset by $0.36 million – which is 9% of the initial construction cost. This is a good reason for developers to be extremely interested in the nuances of the tax code!

Table 2 Tax Benefits of Depreciating an Asset With Initial Book Value of $4 Million and a Life of Ten Years ($ millions)

Year	Straight Line		Double Declining Balance		Double Declining Balance Reverting to Straight Line	
	Depreciation	Tax Benefit	Depreciation	Tax Benefit	Depreciation	Tax Benefit
1	$0.3 million	$0.120	$0.800	$0.320	$0.800	$0.320
2	$0.3 million	$0.120	$0.640	$0.256	$0.640	$0.256
3	$0.3 million	$0.120	$0.512	$0.205	$0.512	$0.205
4	$0.3 million	$0.120	$0.410	$0.164	$0.410	$0.164
5	$0.3 million	$0.120	$0.328	$0.131	$0.328	$0.131
6	$0.3 million	$0.120	$0.262	$0.105	$0.2621	$0.105
7	$0.3 million	$0.120	$0.210	$0.084	$0.2621	$0.105
8	$0.3 million	$0.120	$0.168	$0.067	$0.2621	$0.105
9	$0.3 million	$0.120	$0.134	$0.054	$0.2621	$0.105
10	$0.3 million	$0.120	$0.107	$0.043	$0.2621	$0.105
Total		$1.200	$0.086	$1.428		$1.600
NPV		$0.737		$1.023		$1.096

Now think about what would happen if the owner didn't have any taxable income. In that case, there would be no tax benefit from the depreciation, however it was calculated. The owner in such a situation might consider selling the asset to someone who could use those tax benefits – and who would allow the seller to lease the building on a long-term lease. Or, what is almost the same thing, the business would simply lease assets from companies that could get tax benefits from ownership and pass them on to the business in the form of low lease rates.

Let's return to the situation where the owner does have plenty of taxable income and can therefore take advantage of the tax benefits. In year 11, a change may be needed, as there will be no further tax benefits from the fully depreciated asset. The owner's taxes will therefore increase by $120,000 per year. This conceivably would be a good time to sell the asset (if it still has useful remaining life), so that someone else could capture tax benefits from a new process of depreciation.

Depreciation is especially important in real estate, as buildings are very long-lived assets that can increase in value if they are well-maintained or in a good location. If the building's cost can be covered by a mortgage, and if rents are sufficient to cover the mortgage payments and upkeep, then the tax advantages of depreciation could be very important to the financial structure of the project. For example, consider a skyscraper that is constructed at a cost of $450 million that has a salvage value of $50 million. If this building is depreciated using straight line depreciation over 40 years, then the depreciation would be $10 million per year and the tax advantage would be $4 million per year for a business with an effective tax rate of 40%. For someone with a discount rate of 10%, this tax benefit would be worth approximately $4 million/0.10 = $40 million.

If a capital asset is sold, then the difference between the sale price and the depreciated basis for the asset will be treated by accountants and the tax collectors as a **capital gain or loss**.

(Eq. 12) Capital Gain = Sale Price – Book Value

The accounting logic (or perhaps we should say the "accounting fiction") is that the value of the asset actually equals the depreciated basis that is reported to the tax collector and to shareholders up until the moment that the asset is sold. At the time of the sale, there is an instantaneous gain or loss in the value of the asset. Some assets, like automobiles, lose much of their value as soon as they are first put into service, so their depreciated basis will at first overstate the value of the asset. Other assets, notably real estate, not only depreciate less rapidly than assumed by the accounting rules, they are very likely to increase in value. Capital gains are typically taxed at a lower rate than income, so there is an advantage to companies and to individuals to receive cash payments as capital gains rather than as income.

Depreciation therefore is an accounting technique that protects some cash receipts from being taxed as income, while allowing taxes on any increases in value to be deferred until the asset is sold – and then taxing the capital gain at a lower rate! For owners of real estate, who expect their properties to increase in value, the tax rules related to depreciation and capital gains can be extremely favorable! From the public's perspective, the favorable treatment of real estate may also be quite acceptable, either as a way to reduce the costs of providing rental housing or as a way to help attract investments in retail outlets, office buildings, and other businesses that provide jobs and increase regional income.

Depreciation, while important, is something that is typically governed by accounting standards and federal tax regulation, both of which are slow to change and neither of which is susceptible to manipulation by local governments. Local governments therefore will be much more interested in the tax incentives that they control, namely local sales taxes, local income taxes, and real estate taxes.

Local taxes can be used in two ways:

- Specific taxes can be targeted to specific projects, e.g. sales taxes are sometimes used to finance transit projects.
- Real estate taxes can be reduced in order to make certain kinds of development more attractive, e.g. a city might encourage a manufacturer to locate a new plant by offering a reduced real estate tax rate for a period of years.

As was the case with depreciation and capital gains, the key consideration will be the effect of changes in tax laws on the cash flows associated with potential projects. If taxes are increased in order to finance public projects, then the public authorities must satisfy voters that the benefits of the proposed projects will be worth the increase in taxes. Specific proposals will often be presented to voters in a referendum, and voters may decide whether or not to spend extra money to build schools, improve roads, construct bridges, or create a new sewerage system.

Tax deals designed to attract businesses or to encourage specific projects may or may not receive the same scrutiny, as it usually proves easier politically to provide tax breaks that promote a project than it is to raise taxes on voters in order to finance a project. Nevertheless, it is well worth considering whether the alleged economic benefits justify the tax concessions that are used to attract development. Large national and international corporations will be able to encourage competition among regions and even countries, and they may eventually locate where local governments are willing to forego almost all taxation. It is possible – but not desirable - for local governments to forego tax revenues even though they commit to considerable spending for local infrastructure and services related to the new project (e.g. new roads, increased road maintenance, increased police protection, and general increases in education and other public services related to population growth).

Land Use Regulations

Local governments can also affect development through zoning and other regulations related to land use. If there are no limitations on land use, then developers may attempt to construct whatever provides them the greatest financial benefits. In some locations, this could be high-rise office buildings, while in other locations it might be new factories, a fast-food restaurant, or a recycling facility. Some cities make little or no attempt to limit development, allowing developers and land owners to do whatever they like with their land. In other locations, cities have elaborate plans that are used to guide development, and different parts of the city are zoned for different types of development.

Rationales for zoning include the separation of incompatible land uses, protection of scenic or environmentally sensitive areas, preserving land for special uses, and limiting the density of development to preserve the aesthetic or social character of the community. Opponents of zoning may have little faith in the ability of planners to direct development, they may believe that land owners should be able to do whatever they want with their land, or they may fear that greed, politics, and corruption will overrule fairness and common sense in planning. It is beyond the scope of this essay to delve into the pros and cons for planning and zoning, but it is central to our text to understand how zoning and other land use regulations affect the value of land and the potential for projects.

Since zoning can restrict the types of development that are allowed, it may prevent developers from building projects that they believe would be the most profitable. If so, then landowners or developers may seek a zoning variance that would allow them to undertake a different kind of project or a larger project. To secure a variance, whoever owns a plot of land may apply to the zoning board or whatever authority controls the land use regulations. If the land is re-zoned, then the more profitable project will be feasible, so the value of the land increases – perhaps very considerably. If land formerly zoned for single-family houses on 1 acre lots is re-zoned to allow apartment buildings, then land values could double or triple. If the land is re-zoned to allow construction of a large mall, then a few acres of land could suddenly rise in value from less than $100,000 per acre to $10 million or more!

When public planning commissions must vote on whether or not to re-zone land for more lucrative developments, there are obvious opportunities for political maneuvering, shady financial deals, double-dealing, and worse. Consider the scandal associated with a proposal to create a colossal sports and gambling complex in the site of a defunct racecourse situated within a large public park near Dublin, Ireland. The gambling scheme failed, despite much political maneuvering, transfers of envelopes of cash, lavish "corporate hospitality" at football matches, and offers of lucrative consulting contracts to political opponents. The scheme failed because of well-organized opposition that gathered 20,000 signatures from local residents against the introduction of gambling into one of the wealthier suburbs of Dublin. The promoters of the scheme had no option but to sell the old racecourse to another group, developers Flynn and O'Flaherty, who followed local government's incentives that promoted construction of high-density housing in locations served by Dublin's public transit system:

> *Flynn and O'Flaherty set about having the racecourse rezoned for residential development ... [and] secured full planning permission ... to roll out more than 2,300 new homes, along with an 18-acre public park and other community facilities including bars, restaurants, shops, crèches and a primary school. In May 2004, when the first phase was launched, Flynn and O'Flaherty grossed 110 million pounds in one day and stood to make vastly more money from the Phoenix Park Racecourse than the bookies ever did in 90 years of racing,' as Jack Fagan put it in the times."*[2]

Among the promoters of the original gambling scheme was Mr. Bertie Ahern, the Taoiseach of Ireland (Taoiseach is the title for the prime minister of Ireland). After much public haggling over his involvement, which apparently included more than $1 million in dubious transfers to and from his bank accounts, Mr. Ahern announced his resignation on April 2, 2008, while still denying that he had received a corrupt payment.

---

[2] Frank McDonald and Kathy Sheridan, **The Builders,** Penguin Press, Dublin 2008pp. 54-58

Zoning restrictions may also limit the size or height of buildings or the location of buildings on a site. Washington DC and Paris are among the cities that do not allow skyscrapers to be built in the center of the city in order to preserve the architectural integrity and prominence of their historic structures. New York City, after allowing hundreds of skyscrapers to be built in the lower portions of Manhattan Island, found that these immense structures could transform streets into dark canyons. In order to prevent further crowding of the airspace, the city instituted zoning laws in 1916 that allowed buildings to be constructed right out to the sidewalks, but required **setbacks** as the buildings rose higher:

> *"On 25 percent of the plot, buildings could rise as high as technology and the will of the developer were willing and able to take them. The law divided the city into zones that allowed buildings to rise a different multiple of the width of the street before requiring a setback. In areas zoned for the most intensive use, ... buildings could rise straight up from the building line two and a half times the width of the street before setbacks were required. ... In some specialty zoned retail districts buildings could rise only one and a quarter times the street width, but in most of Manhattan buildings could rise one and a half times the width of the street."* [3]

The result was that the city could benefit from the density of skyscrapers without having huge buildings block all of the light and air from reaching the street levels. Also, by allowing towers of any size on a quarter of the lot, this zoning allowed very tall buildings. The setbacks required by this zoning law were reflected in the design of the Empire State Building, the first 100-story building in the world.

Figure 4 Typical Sidewalk View in Downtown Vancouver, British Columbia: despite the prevalence of high-rise buildings, downtown Vancouver enthralls visitors with numerous tiny parks, sculptures, and waterfalls, which makes walking to and along the waterfront a delightful way to pass a couple of hours.

Another approach to zoning is to limit the **floor area ratio (FAR)**, which is the ratio of the sum of the total floor area of a building to the size of the site. A floor area ratio of 14, for example, would allow a 14-story building that covered the entire lot or a 28-story building that covered half of the site. The same FAR would allow a mixture of buildings on a site. For example, with the same FAR of 14, it would be possible to build a 40-story building on a quarter of the lot, a 6 story building on two thirds of the site, while leaving the remaining portion of the site as open space: 40 (1/4) + 6 (2/3) = 14.

---

[3] John Tauranac, **The Empire State Building: The Making of a Landmark**, St. Martin's Griffin, New York, pp. 56-57

In suburban and rural locations, zoning is frequently used to limit the density of development by specifying a **minimum lot size** and limiting development to single-family residences for large portions of the region. The minimum lot size in a suburban location may be a quarter of an acre or less, while the minimum lot size in a rural location is sometimes set as high as five acres. These restrictions are certainly effective in limiting the total number of housing units, but they may also lead to sprawl, i.e. to the conversion of large portions of farmland and open space to residential uses.

A different approach to limiting density is known as **cluster zoning**, in which the density of development is still stated as the number of units per area, but the housing units (or other structures) can be clustered in a small portion of the site, leaving large portions of the site as open space. Cluster zoning can be very effective in minimizing the environmental impact of development, not only because it leaves much of the site as open space, but because it allows utilities and transportation services to be more efficient, as people and houses will be closer together.

Building codes and Other Safety Standards

Governments establish building codes in order to reduce the risks associated with structural failures, accidents, fire, floods, earthquakes, and other natural or man-made hazards. Minimum standards for construction are needed to protect people from unsafe or unsanitary conditions in their buildings. As building technologies change – and as accidents or natural disasters identify problems – building codes may be updated.

Examples of regulations that are designed to reduce the risks of natural hazards would include the following:

- Design standards and inspection requirements for dams.
- Requiring brick firewalls between attached structures in order to limit the possible spread of fire.
- Requiring sprinkler systems and fire escapes for schools, apartment buildings, and office buildings.
- Higher standards for structures constructed in earthquake zones.
- Prohibitions regarding the use of lead pipes, asbestos insulation, and other materials that are known to cause health problems.

Regulations designed to improve safety would include:

- The use of nets and other precautions when constructing bridges and high rise buildings.
- Requiring railings for steps on residential housing.
- Requiring numerous safeguards for nuclear power plants.

Environmental Regulations and Restrictions

Environmental regulations may affect what, how, where and when something is built. Examples include:

- Minimum setbacks from rivers, ponds, lakes, and wetlands
- Restrictions on filling wetlands
- Restrictions on development in flood plains
- Designation of areas as wilderness

Minimum setbacks from water are desirable for several reasons. Maintaining a natural buffer will help to limit water pollution resulting from runoff of rain water and allow a passage way for wildlife (Figure 5). In cities, pedestrian access to waterways and bodies of water is an important consideration, and many cities have developed linear parks along rivers, lakes or the ocean.

Figure 5 The landscaping for this mall in Front Royal, Virginia features planting of trees and shrubs along the roads and parking lots and preservation of wetlands between the mall and the main access road from Interstate 66. Section 175-53.1 of the zoning code provides guidelines for development of "Highway Corridors" along the major entrances into the town and "strives to ensure that such development is compatible in use, appearance, and functional operations with the Town's economic development policies and action strategies."

Restrictions on filling wetlands recognize the importance of wetlands for many species of animals and plants as well as the ability of wetlands to act as a natural buffer for holding water following heavy rains or snow melt. When building roads or other transportation routes, it will at times be necessary to cross wetlands, but there may be routes that could be chosen that would limit the area of wetlands that would be filled in and there may be construction techniques (e.g. use of bridges rather than a causeway) that limit the disruption to the wetland. It may also be possible to create new wetlands alongside the highways, so that the net impact of the highway construction could be mitigated in some locations.

Some states have created a system that requires highway construction and other development projects to cover the costs of preserving and protecting wetlands. When new roads are constructed, the state department of transportation is required to a) limit the destruction of wetlands, if possible, b) to create new wetlands, if feasible, and c) to make a payment into a special fund for any wetlands that are destroyed. A separate agency uses the money from this special fund to a) acquire and preserve wetlands elsewhere in the state and b) to create new wetlands in locations where that is desirable. This process can be structured so that the area of wetlands that is preserved is much greater than the amount that is destroyed.

Individual landowners can also take action to limit future development by putting conservation easements on their property in order to limit future development. Easements may prohibit any development, or they may allow limited development such as the construction of a single house on a specified portion of the site. They may allow activities such as agriculture, forestry, sports (e.g. hiking or skiing), or they may require the land to remain in or revert to a wild state. An easement is a legal document. When an owner puts a restrictive easement on property, the value of that property will decline. The decline in value can, in some instances, be considered to be a charitable donation, as some legislation encourages landowners to provide easements aimed at preserving open space or protecting wildlife. The charitable donation is an amount that can be deducted from income when calculating how much is owed in income tax, so there can be a financial benefit to an individual or a company for placing an easement on a property. The amount of the deduction must be verified by qualified assessors who determine the value of the property with and without the easement, based upon recent sales of similar properties.

Summary

Income tax is based upon net income, the difference between income and expense. Income taxes may be levied by local, state, or federal governments. Local and state taxes may be deductible from income when calculating federal income tax. The effective tax rate combines the effects of local, state, and federal taxes. The after-tax MARR equals the pre-tax MARR multiplied by one minus the effective tax rate. Since companies and investors are concerned with after-tax cash flows, their decisions concerning where to invest, when to invest, and how much to invest may be affected by tax policy. For example, governments may allow tax credits to promote certain types of infrastructure investments.

Since capital investment is not an expense, investment does not affect profit or income tax. However, capital assets decline in value over their lifetime, and the depreciation in the value of capital assets can be treated as an expense. Depreciation is a non-cash expense that represents the actual or assumed decline in value of assets other than land, which cannot be depreciated).

Depreciation, although it is a non-cash expense, will be deductible from income. Since depreciation can reduce income tax payments, the rules governing depreciation will affect the perceived value of a project to investors. Tax laws and accounting rules determine the methods that can be used to depreciate an asset. One commonly-used method is straight line depreciation, which assumes that a capital asset declines in value at a uniform rate over its life. Other commonly-used methods include double-declining balance and asset class depreciation, both of which allow higher rates of depreciation during the early years of an asset's life. Accelerated depreciation results in lower profits, which may seem undesirable, but it also results in lower taxes and an increase in the net present value of a project. Public policy may therefore allow accelerated depreciation as a means of encouraging certain types of investments.

Zoning limits the type of development that can be undertaken in a city or town. Certain areas may be zoned for residential, while others are zoned for commercial or industrial. Zoning may require a minimum lot size, designate how close a building can be to property boundaries, or limit the maximum extent of development, e.g. by specifying a maximum floor area ratio (FAR). Since the value of land is largely dependent upon the potential for development and the character of the surrounding areas, zoning may either enhance or depress land values. For example prohibiting obnoxious development within a residential neighborhood may increase the value of the homes in that area, but prohibiting more intensive development (e.g. apartment buildings or smaller lot sizes) may depress nearby land values.

Developers often seek zoning variances in order to initiate more intensive or what they believe to be better (and certainly more profitable) uses of the land. Where re-zoning would cause a dramatic increase in value – but equally dramatic changes in a neighborhood – there is potential for great local political conflict and controversy.

Land owners may choose to accept restrictions on their own property by creating an easement. For example, an easement could allow a right-of-way (e.g. for power lines or for access to other properties), allow public access, or allow or prohibit certain uses of the land. Conservation easements are commonly used to promote preservation of open space. Such easements may prohibit further development or allow only agricultural or forestry activities. If landowners donate an easement on their property to a qualified charitable organization, then they can treat the assessed reduction in the property's value as a tax deduction.

Building codes define what materials and methods can be used in construction, and they may mandate better construction techniques and materials in areas prone to natural hazards, especially earthquakes or flooding. Safety standards may be part of building codes. For specialized facilities, such as nuclear power plants, safety standards may require great care (and great expense) in design, construction, and operation. As technology advances, the logic underlying building codes and safety codes may need to be revised to allow cheaper or more effective methods and designs (e.g. plastic pipes for plumbing). As knowledge of environmental impacts and risks increases, building codes may need to be revised to avoid dangers that were previously unknown (e.g. the use of asbestos for insulation is banned because this material is linked to lung cancer).

Environmental regulations may affect what, how, where and when something is built. Examples include:

- Minimum setbacks from rivers, ponds, lakes, and wetlands
- Restrictions on filling wetlands
- Restrictions on development in flood plains
- Designation of areas as wilderness

Legislation or regulation may require infrastructure projects to mitigate their effects on the environment. For example, transportation projects that disrupt wetlands may be required to pay money into a special fund that is used to protect or to create other wetlands.

Figure 6 Pier A Redevelopment Project, Battery Park, Manhattan
The City of New York and the Battery Park City Authority are working with the NY City Economic Development Corporation to rehabilitate this historic Pier, which was constructed in the 1880s. The goals are enhance access to the waterfront, create jobs, and produce economic benefits for the city.

Dealing With Risks and Uncertainties

Our fundamental need is not the elucidation of the mysterious, but an appreciation of the significance of the obvious.[1]

Introduction

When infrastructure projects are planned, especially for new types of infrastructure, there will be many uncertainties concerning operating performance, demand, profitability, environmental impacts, safety, and other social and economic impacts. One approach to dealing with such uncertainties is to include contingencies when estimating the time and costs that will be incurred, thereby having a buffer to deal with problems that may emerge. Perhaps past experience with similar projects has reduced the uncertainty about a particular type of infrastructure project, so that project managers have some confidence about what kinds of problems are likely to emerge and what can be done to deal with them. Once a project has been completed, it will still be possible to make some adjustments to improve performance and deal with unexpected side effects. However, it will always be wise to consider risks and uncertainties while evaluating alternative project designs, rather than waiting to find out what problems emerge after a new system is open for business.

Risk and uncertainty are sometimes used interchangeably, but there are some differences worth mentioning. The term "risk" implies that something bad could happen, which might involve something physically bad (a building collapses) or something less tangible (mortgage markets dry up and financing for the project cannot be obtained at a reasonable interest rate). The term "uncertainty" does not have the connotation of something that is bad; instead, it refers to the inability to predict exactly what will happen. Interest rates could go up or down; demand for apartments could go up or down; the costs of gasoline or building supplies could rise rapidly, slowly, or oscillate.

Common risks associated with any major project will include the following:

- Construction risks: it may not be possible to construct the project as planned, on budget, within the original time schedule. Storms could cause extensive damage, and unexpected geotechnical problems could set back a project many months or years.
- Financial risks: it may not be possible to obtain sufficient funding for a project; interest rates on bonds, mortgages or loans may be much higher than anticipated; investors may demand a larger share of the company ownership or pull out of the project altogether; general economic conditions may decline sharply while the project is underway, possibly making it impossible to continue borrowing money to complete a project.
- International risks: if a project is undertaken in another country, exchange rates could change, thereby upsetting key assumptions about the value of cash flows. In some locations, there could be a risk that local governments will change the regulations governing the project, attempt to take over the project or fail to follow through on commitments made to support the project.
- Infrastructure safety: there will be risks of accidents both during construction and over the life of the project. These risks could affect construction workers, operators, users, or abutters. Avoiding infrastructure failure and reducing risks associated with operations and maintenance should be fundamental goals of the engineering design process.
- Demand risks: the revenue projected for a project may fail to materialize because of a lack of demand or overly optimistic assumptions about pricing.
- Operating costs: the project might not work as well as planned; operations might be more costly and service might not be as good as expected.

---

[1] John A. Droege, **Freight Terminals and Trains**, McGraw Hill, New York and London, 1925, p.13

Some of these risks, such as the risk of being unable to complete the project on time, can be reduced by allowing buffer time in the schedule and a significant amount for unexpected contingencies in the budget. Risks related to storms can be reduced by developing emergency plans and by taking precautions at the building site. Safety risks can be mitigated by testing materials, ensuring good design standards, and careful monitoring of all construction processes. Risks related to demand and operations can be handled by undertaking detailed studies of how the system will work and how potential users will respond to the system. Uncertainties in demand relate to the size and staging of projects and to the design of the project. Understanding how potential customers might respond to new infrastructure is therefore an essential aspect of project design.

Example: Dealing with Risks and Uncertainty in a Toll Road Project

Suppose a city and a company have entered a public/private partnership to construct a new toll road. The city has allowed the company to charge a fixed toll during the first years of operation, with some ability to raise tolls every few years thereafter. The city recognizes that demand will be lower in the first few years of operation, so it has agreed to provide a subsidy to supplement toll revenues for a period of several years. The city has also agreed to guarantee the interest rate on the bonds issued by the company to pay for the construction costs. The company was created to construct and operate the toll road; its future depends upon the financial success of the project. What are the uncertainties and risks in this situation and what are the key issues to negotiate?

The major uncertainties are:

- The cost of construction, which will determine the annual interest rate on bonds
- The time required for construction, which will determine the point at which toll revenues begin
- The demand for the road

The major risks are:

- If the company does not receive enough revenue from tolls and subsidies to cover the interest costs, it will have to declare bankruptcy and the city will have to take over the project.
- If the city has to pay too much for subsidies, then it will have to cut back on other projects or raise taxes, which will cause a political uproar.
- If tolls are set too high, and if the private company becomes very profitable, there could be a political backlash against the project and the ability for the city to undertake similar projects in the future would be jeopardized.

The key issues to negotiate are:

- Initial levels for tolls: if set too low, public subsidies will be too high; if set too high, public outrage could be a problem.
- Escalation for tolls: if tolls are allowed to rise sooner, rather than later, then the private company may be able to borrow additional funds to cover losses in the initial years. The contract could require the company to establish a fund or a line of credit that would be sufficient to cover a shortfall of toll revenues during the first few years.
- Nature of the subsidy: if a maximum is established, then the company bears more risk related to the uncertainty in demand; if no maximum is established, then the city bears all of the risk related to the uncertainty in demand

If the city and the private company are careful in structuring the deal, then they should be able to reduce the risks that they each face. If one party fails to recognize the major uncertainties, then they are likely to end up with more of the risks.

Using Analysis to Understand Risks

More detailed methodologies can be used to gain a greater appreciation of the risks and uncertainties associated with a project, including the following:

- Sensitivity analysis: sensitivity analysis can be used to find out which aspects of a project are likely to be most critical.
- Modelling performance: infrastructure-based systems can be evaluated in terms of many measures, including cost, service, capacity, and safety, and projects may be aimed at improving any or all of these aspects of performance. Given a well-structured model, it will be possible to investigate how a proposed project might improve performance and how the improvement in performance might affect demand or profitability of the system. Simulation or analytical models can be used to study many aspects of performance, including cost, service, and capacity.
- Scenario analysis: the basic idea of scenario analysis is that potential projects should be considered in the context of various possible visions of the future, taking into account the political and social context as well as engineering, economic, and financial issues. In deciding whether or not to proceed with a project or how best to proceed with it, it will be very helpful to consider multiple futures and to consider the different risks that might be encountered in each of them.
- Probabilistic risk assessment: this is a technique that can be used in developing strategies to improve the safety of infrastructure systems by assigning probabilities to various types of accidents or incidents that might be encountered and also considering the range of potential consequences of each type of accident or incident.
- Performance-based technology scanning: new technologies may reduce cost, improve operations, increase capacity, enhance safety, or improve some other aspect of performance. Whether or not new technologies are worth pursuing will depend upon how potential customers or users will respond to the improvements in performance provided by the new technologies.

Using Sensitivity Analysis to Study Uncertainties and Risks

It takes some thought and considerable care to conduct an intelligent sensitivity analysis and to interpret and present the results of such an analysis. A good sensitivity analysis seeks to determine whether the success of the project is in doubt, given the uncertainty in the various assumptions that must be made. By varying one variable at a time along a range of possible values, it is possible to calculate the effect on return on investment (ROI), net present value (NPV) or any other measure of performance. By varying two or more variables, it is possible to see more complicated interactions. The conclusions of a sensitivity analysis might be stated as follows:

- As long as construction costs are no more than 20% of budget and net revenues are at least 60% of what we expect, the NPV of the project will be positive (a very strong statement that this would be a good project).
- If construction costs are 10% over budget or if incomes is only 90% of what we anticipate, then we will be unable to cover the interest payments on our mortgage (a very clear statement that this is a risky project)

Sensitivity analysis can be used to understand performance under various assumptions about the key factors related to project success. By looking at many possible combinations of factors, it is possible to determine if there is a risk that the project will fail, given the uncertainty in the assumptions related to supply and demand. However, sensitivity analysis alone cannot provide much insight into the underlying engineering, economic, political or financial problems that will affect the success or failure of a project.

Modeling Performance: Simulation and Analytical Models

Greater insights into risks and uncertainties associated with a project can be gained by creating a model or models to investigate how well the project will perform under various sets of assumptions. Models can be simple or complex. A simple probabilistic model could be run dozens or hundreds of times to determine what the most likely performance will be and the chance that performance will be unacceptable. Complex engineering-based models can examine fundamental issues related to design, maintenance, and operations. Planners can use simple models to investigate a wide range of design options, then use more detailed models to study a few promising options in much greater detail.

Whatever the degree of complexity, models can be used to address three sets of questions related to project evaluation:

1. How will system capabilities or performance change if the project is implemented?
2. If system capabilities or performance change, what will happen to demand?
3. How will uncertainties affect the success of the project?

The first set of questions refers to the engineering factors that underlie performance. How does the system currently function? What are the most important relationships? What are the key performance measures? What are the critical factors that affect costs, capacity, or service quality? How will the proposed project affect these relationships, performance measures, and factors? These are engineering questions that can be addressed with models appropriate to the system. For example, putting a portion of a light rail transit system into a tunnel can save travel time by eliminating delays at grade crossings. Modeling the operation of the system may show that the travel time savings would be 4 minutes, reducing the average time from 44 to 40 minutes. Another example might be an analysis of various designs for wind farms to determine which will be the most cost effective in producing electricity.

The second set of questions concerns what might happen after the project is implemented – what happens next? So what if the average travel time on the transit system is reduced by 10% or if the unit cost of producing electricity with wind turbines is reduced by 10%? Will most people still prefer to commute in their own cars? Will the cost of green electricity be competitive with the cost of electricity from efficient gas-fired plants? The changes in performance may or may not result in any more people using the light rail system or in any increase in demand for electricity produced by wind turbines. To determine the effect on demand, it would be necessary to consider the factors that affect the demand for transit and the willingness of customers to choose to use green energy even if is somewhat more costly.

The third set of questions considers the interaction among the key factors that determine the success of the project, which will include variables related to supply and demand as well as variables related to the financing of the project. What happens if costs are much greater than expected? What happens if performance does not improve as much as expected? What happens if resource prices are much higher than anticipated or if new technologies reduce costs of competing projects?

Probabilistic Risk Assessment

Probabilistic risk assessment is a structured methodology for understanding and improving the safety of an infrastructure system. This methodology defines risk as the product of two factors: the **probability of an accident** and the **expected consequences** if an accident occurs.[2] Global risks associated with a particular system can be estimated by summing over all types of accidents and all types of consequences. Investments to reduce risk can be compared by considering the ratio of the reduction in risk to the cost of achieving that reduction.

This methodology address the two aspects of risk assessment that will be critical for many projects. First, what is the most effective way to reduce the risks that affect infrastructure safety? For example, what is the best way to reduce highway accidents or the best way to reduce the risks of floods? Second, how can the risks associated with the safety of a project be best communicated to the public? For example, how does an energy company deal with public perceptions of the risks of nuclear energy? Both aspects of risk assessment relate to the way that people and society perceive risks and decide what risks are acceptable.

---

[2] Probabilistic risk assessment applies to studies related to safety, e.g. the likelihood of accidents that disrupt service and result in property damage, injuries, or fatalities. As mentioned above in this essay, there are many other kinds of risk to be considered, such as the risks associated with financial matters (currency exchange, interest rates) or demand (will demand meet initial projections?)

Clearly, individuals are quite willing to accept substantial risks in their everyday activities. Some people go skiing, a few go sky-diving, kids ride down hills on their bikes, and many people like to ride their motorcycles. Despite the vast number of broken limbs associated with these activities, many people do seem quite willing to buy and use their skis, bikes, and motorcycles, even though many more prefer snowshoeing, walking, and jogging as less risky activities that are more clearly attached to the ground. In general, people may at times decide to avoid activities that they view as too risky, while at other times they seek out activities where the excitement or the view or the sense of achievement justifies whatever risks are encountered.

Collectively, people also are able to decide what kinds of risks are acceptable to society. In some cases, societies not only condone, but promote activities that are well-known to be risky. Since more than 30,000 people are killed annually on highways in the United States, it is clear that driving is risky. Most of us know of someone who was killed or severely injured in an automobile accident, and anyone who drives extensively is likely to be able to recount several close calls. Yet we do not limit access to highways; we do not post speeds of 30 mph to avoid high-impact collisions; we seldom shut down the highways in snow storms; and we do not require construction of automobiles that resemble tanks in their ability to survive collisions. As a society, we do not "do everything possible to save just one human life." As individuals, we each continue to drive, because the benefits of personal mobility outweigh whatever risks we perceive in driving. Quite possibly many of us refuse to believe that we ourselves are at risk, just as many of us believe that the cost of driving is free, so long as there is gasoline in the tank. On the other hand, we do impose speed limits; we do have design standards for highways and vehicles; we do punish drunken drivers; and we do require seatbelts. Some actions have been deemed worthwhile to pursue, while others have not.

Whether or not the logic is stated explicitly by pubic officials, it is useful for us to understand how analysis can help determine what risks are acceptable and what values can be used to determine the costs and benefits of changes in safety. The three key types of factors are as follows:

- Probability of an accident (P)
- Consequences of an accident (C_i)
- Relative importance of the consequences (W_i)

Risk can be defined as the sum of the expected consequences of an accident:

(Eq. 1) $$\text{Risk} = P \, \Sigma(C_i * W_i)$$

If the weights (W_i) are expressed in monetary terms, then changes in risk can be used in benefit/cost analysis. Given the importance of reducing risks and public concerns with any major accident involving infrastructure, it is worth spending some thought on what it really means to reduce the expected consequences of an accident. The direct medical costs associated with an accident can be estimated based upon past history, but what about serious injuries or fatalities? There are many approaches to this issue. A strictly economic approach might consider the average lifetime earnings of an individual killed in an actual accident, but it is impossible to say who would have been involved in an accident that is prevented. A possible solution to this problem would be to consider the age and income potential of the people using the facility. For example, if the typical user is 35 years old, earning $50,000 per year for 30 more years, then their future earnings would be more than $1 million. This approach is fraught with difficulties: "how can we put a value on a human life" is a common refrain after any tragic accident, especially one involving children.

Another approach advocated by economists such as Ken Small would look not at the cost of a fatality but at the value of reducing the risks shared by all users. In trying to improve highway safety, this approach has the merit of focusing on a benefit that is shared by everyone using the highways, rather than directing attention to an unknown victim of an unspecified accident. With this approach, the basic question is what would an individual pay to achieve a small reduction in risk? This question can be answered through detailed interviews with a cross-section of users. Studies have found that people in the United States would be willing to pay a few dollars for a small reduction in the probability that they would die in a transportation accident. If individuals are willing to pay on the order of $2.50 to reduce the odds of a fatal accident by 1 in a million, then it is logical to say that society should be willing to pay on the order of

$2.50 million to avoid a single fatality in an accident. Instead of putting a value on a human life, this approach puts a value on the reduction in risk that is provided to every traveler.

Another approach would be to consider the damages awarded in lawsuits involving accidental death. And a much different approach might just consider the need for a reasonable number to use in estimating benefits and costs for public investments and public policy. If an extremely high value is placed on expected fatalities, then it will be possible to justify extreme measures to improve safety. If no consideration is given to safety, then it will be possible to construct facilities that are quite unsafe. A society, through political and legal processes, will determine acceptable levels of risk and how much is worth investing in order to reduce risk.

Example: Which Automobile Safety Features Are Most Worth Pursuing?

Probabilistic risk assessment can be used to investigate the effectiveness of various strategies for reducing fatalities resulting from highway accidents. If the probability of a fatal accident is 1 in 100 million miles, and if the average car is driven 10,000 miles per year, then the probability of a car being involved in a fatal accident would be 1 in 10 thousand per year. Let's assume that the benefit to society of reducing risks so as to eliminate one fatality is $2.5 million. The expected costs incurred from an average car being involved in a fatal accident would therefore be $250 (0.0001 fatal accidents per average car per year multiplied by $2.5 million per fatality = $250/year). Consider three possible changes to reduce risk, each of which would increase the initial cost of the car:

- Require seat belts: an investment of $100 per car reduces the probability of a fatality by 50% (to 0.00005 per year); risk would decline from $250 per year to $125/year, a benefit of $125 per year.
- Airbags for driver and passenger: an investment of $500 per car further reduces risk of fatality by 80% beyond what is possible with seat belts (to 0.0000125 per year); risk would drop from $125 per year to $25 per year, a benefit of $100.
- Automatic collision avoidance system: an investment of $20,000 per car further reduces risk of fatality by 10% beyond what is possible with seat belts and airbags (to 0.0000113); risk would drop from $25 per year to $22.50 per year, a savings of $2.50 per year.

To compare the risk reduction benefits to the costs of the systems, the additional investment costs must be converted to annual costs. With typical assumptions about financing automobile usage and financing, the annual costs of these three options would be approximately $12 for seat belts, $60 for airbags, and $2,400 for the collision avoidance system.[3] The reductions in annual risk per automobile would be $125 for the seatbelts, $100 for the airbags, and $10 for the automatic collision avoidance system. The seat belt and airbags are clearly justifiable, as the expected reduction in risk is far greater than the annual cost. The hypothetical collision avoidance system, which would cost an additional $2,400 per year, would provide an additional benefit of only $2.50 per year in risk reduction compared to a car equipped with seat belts and airbags. Therefore, it would be better to spend additional money not on automatic collision avoidance systems, but on some other means of reducing risk on the road or in our lives.

Global Risks

Global risk encompasses all of the risks associated with the many types of accidents that could occur within a system. A company or agency should be interested in understanding global risk in order to be able to develop effective programs for mitigating overall risk. Without a clear understanding of overall risk, it is quite likely that too much will be spent to avoid what in fact are minor risks while too little will be spent to avoid what are actually major risks.

Estimating global risk for an infrastructure-based system involves the following steps:

- Identifying the types of accidents that might occur on each portion of the system

---

[3] The cost of each option was converted to an equivalent annual cost assuming a 14 year life for the car and a discount rate of 8% for the owner.

- Estimating the probabilities of such accidents as a function of time or usage
- Estimating the expected consequences of each type of accident
- Calculating the risks associated with each type of accident by multiplying the probability of an accident by the expected consequences of an accident
- Summing risks over all type of accidents

These steps can be carried out at various levels of detail. For a well-established system, estimates of accident probabilities and consequences can be based upon past experience. For new systems, risk estimates will have to be derived from models, comparison with similar systems, or expert opinion.

Different types of risks will have to be considered:

- Risks related to construction
- Risks inherent in normal operations
- Risks related to maintenance and rehabilitation
- Risks related to deteriorating infrastructure
- Risks related to structural failure
- Risks related to natural disasters, such as earthquakes, floods, hurricanes or tornadoes

Some risks are associated with potentially catastrophic accidents that could result in hundreds of fatalities, extensive disruptions in major cities, and extreme property damage. Examples of catastrophic accidents would include the failure of a major bridge, collapse of a building, flooding of a major city, an explosion in a chemical plant, failure of a nuclear power plant, or a plane crash. Many aspects of system design, operating strategies, and inspection and maintenance policies are aimed at avoiding catastrophic accidents, and the probabilities of such accidents are therefore likely to be very small. Nevertheless, because the potential consequences are very large, the risks associated with catastrophic accidents are likely to be an important component of global risk for any system.

Table 1 illustrates some of the major risk factors that might be associated with three hypothetical projects. Using these factors, it is readily possible to estimate the fatalities that could be anticipated over the 100-year lifetime of the projects. The table shows the expected fatalities associated with construction, normal operation, and structural failure. The first project is the construction of a major canal, assuming technology and medical knowledge available at the time of the construction of the Panama Canal. The greatest risks are those associated with the construction process, as a very large work force is necessary and past experience suggests a death rate of 5%. The canal would involve some risk in operation and maintenance, but there are not that many people who would be working at the canal on a daily basis, and there would be minimal risk in the event of a major structural failure. Thus, for the canal, the global risks are primarily associated with the construction, not with operation.

The second project is the construction of a skyscraper, using rough estimates of risks assuming only modest attention to safety during construction. There is some risk inherent in construction, as many workers will be working high off the ground. However, with nets and other safety procedures in place, there should not be many fatalities during construction. Operations of a skyscraper should also be very safe, but it would surely be catastrophic if the building collapsed. The greatest concern for this project is therefore to ensure the structural integrity of the building. The third project is the construction of an urban highway that is expected to serve 100,000 vehicles per day. With this project, the risks in construction can be mitigated through use of proper safety procedures, but there will be a continuing risk associated with traffic accidents. The failure of a portion of the highway would surely attract national attention, but the nature of a highway is such that failure is unlikely to affect more than a few dozen vehicles. The greatest challenge is therefore to improve the safety of the highway.

Table 1 Risk Factors for Three Hypothetical Systems

	Major Canal	Skyscraper	Urban highway
Risk of fatality per construction worker	0.05	0.001	0.001
Construction workers	20,000	1,000	1,000
Risk of fatality per user, normal operation	0.2 per million	0.02 per million	2 per million trips
Users per year	1 million	3 million	30 million
Daily usage	40 ships per day	3,000 occupants	100,000 vehicles/day
Potential fatalities from structural failure	10	2,000	100
Likelihood of failure in 100 years	0.01	0.01	0.01
Expected fatalities			
• Construction (total)	1,000	1	1
• Operations (per year)	0.2/year or 20 in 100 years	0.06/year or 6 in 100 years	60/year or 6,000 in 100 years
• Failure (life of project)	1 in 100 years	20 in 100 years	1 in 100 years

Estimating global risks is an objective exercise that is limited only by the ability to envision the types of accidents that might be expected and the ability to estimate the consequences of those accidents. Once the risks are understood, it is possible to determine which risks are most serious and to consider how best to mitigate those risks. However, it would be a mistake to believe that risks can be dealt with objectively, without regard to the way that people perceive and respond to risk. As discussed in the next section, some risks are perceived to be much worse than others, so that it is necessary to consider perceptions when assessing risks.

Perceived Risks[4]

Estimating "perceived risks" requires more than the assessment of accident probabilities and expected consequences. Adjustments are needed to reflect the perceived importance of certain types of accidents or consequences to various stakeholders, including users and the public. Quantifying perceived risks requires answers to questions such as "Who is at fault?" "Is it a catastrophic accident?" and "Is new technology involved." In principle, it is possible to answer these questions based upon surveys, and researchers have created a framework for understanding how perceptions of risk vary with the circumstances related to particular types of accidents.[5]

One of the first studies of perceived risks found that there is a different trade-off between risks and benefits for activities undertaken voluntarily than for those undertaken involuntarily[6]. People will accept risks from voluntary activities (such as skiing) that are roughly 1000 times as great as they would tolerate from involuntary hazards that provide the same level of benefits[7]. This study concluded that the acceptability of risk from an activity is roughly proportional to the third power of the benefits for that activity.

---

[4] The material in this section is largely based upon a report prepared for the East Japan Railway Company: C.D. Martland, J.M. Sussman, L. Guillaud, and J. Vanzo, "Risk Assessment: Improving Confidence in JR East, Final Report, Research Conducted for the East Japan Railways, MIT Engineering Systems Division, Cambridge, MA, March 2005
[5] For an excellent overview of the fundamental research on risk perception, see Nancy Nighswonger Kraus, "Taxonomic Analysis of Perceived Risk: Modeling the Perceptions of Individuals and Representing Local Hazard Sets", Thesis Ph. D, 1985, University of Pittsburgh.
[6] Starr, C. (1969), "Social benefit versus technological risk", Science, 165, 1232-1238.
[7] Paul Slovic, "Perception of Risk", **Science**, Vol. 236, 17 April 1987.

Additional research on perceived risk identified other factors that affect perceptions of risk. Slovic, Fischhoff and Lichtenstein (1979) compiled from their own and prior studies a list of nine characteristics of risk that might be important in risk ratings by the public: [8]

(1) Whether the risk is assumed voluntarily or involuntarily
(2) Whether the risk of death is immediate or delayed
(3) Whether the risk is known or not to those exposed to it
(4) Whether it is known by science or not
(5) Whether the risk can be controlled by an individual's skill
(6) Whether the risk is new and unfamiliar as opposed to old and familiar
(7) Whether the fatalities are common or catastrophic accident
(8) Whether or not the thought of the risk evokes a feeling of dread
(9) Whether or not there is a risk of a fatal accident.

Slovic, Fischhoff and Lichtenstein found strong intercorrelations among these nine characteristics and suggested that the risk perceptions could be explained in terms of two basic factors (see Figure 1). The factor labeled "dread risk" is defined at its high end by a perceived lack of control, feelings of dread at the nature of an accident, and potential for a catastrophe. Factor 2, labeled ""unknown vs. known", is defined at its high end by hazards judged to be unobservable, unknown, new and delayed in their manifestation of harm. Research has shown that lay people's risk perceptions and attitudes are closely related to the position of hazards within this type of factor space[9]. Most important is the horizontal factor "dread risk" The more a hazard is perceived to be "dreadful", the more people wanted to see its risks reduced, and the more they wanted to see strict regulation employed to achieve the desired reduction in risk.

Figure 1 Reducing Various Characteristics Relevant to Risk Perceptions to Two Key Factors

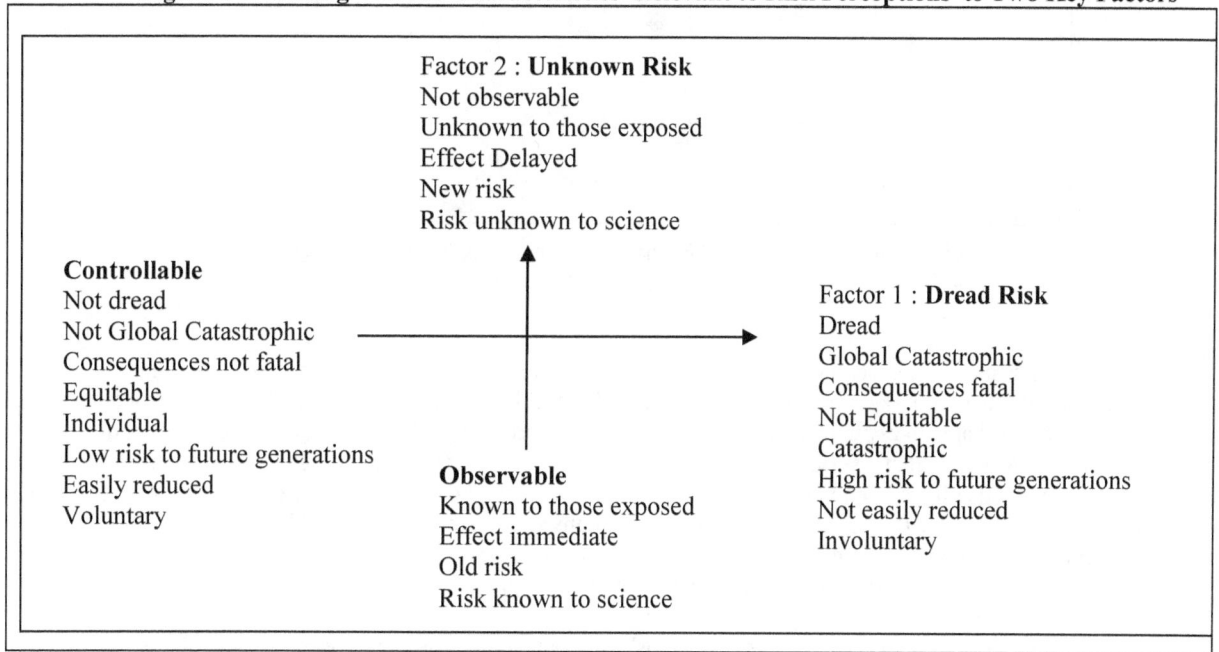

Source : P. Slovic, B. Fischhoff, S. Lichtenstein, "Perilous Progress: Managing the Hazards of Technology", (Westview, Boulder, CO, 1985), p. 91-125.

---

[8] P. Slovic, B. Fischhoff, S. Lichtenstein, "Perilous Progress: Managing the Hazards of Technology", (Westview, Boulder, CO, 1985), p. 91-125
[9] Paul Slovic, "Perception of Risk," **Science**, Vol. 236, 17 April 1987.

Figure 2 shows how people viewed eight hazards, including some related to daily activities, some related to new technology, and some related to infrastructure. The least dreadful risk was that associated with caffeine – we know it isn't all that good for us, but we still have to wake up in the morning! Likewise skiing might get you a broken leg, but that risk is not viewed as dreadful. The most dreadful risks for the people in this survey were those associated with nuclear reactors and radioactive wastes - which goes a long way toward explaining the difficulty of expanding the use of nuclear power in the U.S. and in many other countries. The other scale goes from known to unknown. The risks of automobile accidents are well known because they are so common; we continue to drive despite the risks because we often would rather be somewhere else – to meet with friends, to go to work, to take a trip, or to go shopping. We know the risks, and we accept them. The other end of the scale includes the risks associated with new or unusual technologies. The risks associated with nuclear reactors and nuclear waste are rather high on this scale as well: we don't know how bad the risks really are, but we know we don't like them. DNA technology, which relates to issues such as genetic alterations to increase agricultural yields, is the least known risk and therefore one of the most feared.

Figure 2 Location of Eight Hazards within the Two-Factor Space

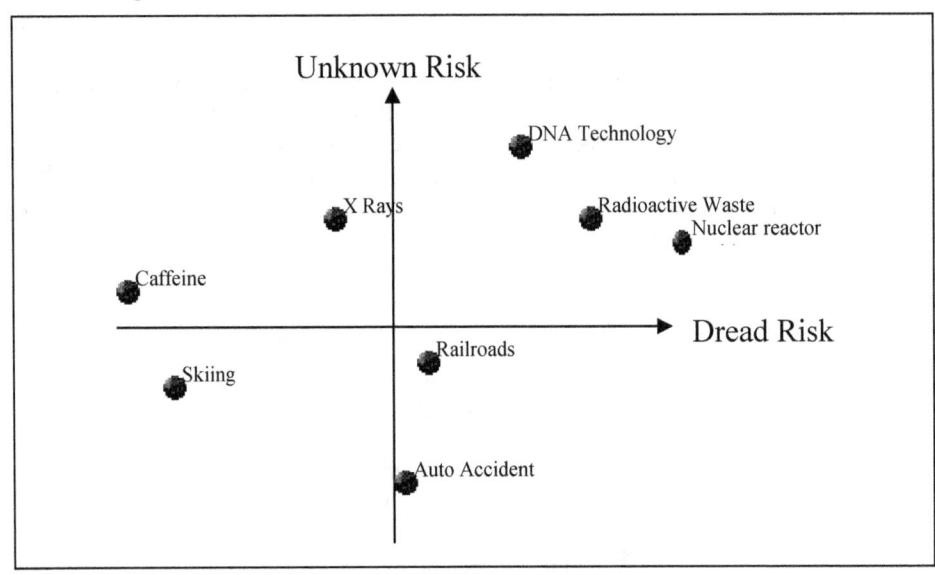

Adapted from: P. Slovic, B. Fischhoff, S. Lichtenstein, in "Perilous Progress: Managing the Hazards of Technology", (Westview, Boulder, CO, 1985) p 91-125.

There are various strategies for dealing with perceived risks. One is to provide more and more safeguards so as to convince the public that the risks are well under control. The problem with this approach is that a rational response may do little to ease what is likely to be in large part an emotional problem. Moreover, the public may just not believe the technicians who claim that everything will work as planned. A second approach is to provide information concerning the plans that have been put in place to deal with incidents, however unlikely they may appear to be. By discussing the risks and how they are being handled, a company or an agency may be able to assuage public fears or to at least provide an impression of competence. A third approach is to involve the public in the design and implementation of a comprehensive and credible risk management plan.

Another, more general approach, is to make sure that risks are put in the proper perspective. None of us live in a risk free environment. We could be involved in a serious automobile accident; we could be hit by a car when crossing the street; we could be hit by lightning or slip on the ice or trip over a chair when rushing to answer the phone. Our house could be hit by a tornado, or we might be caught in a flood. We could be mugged, shot, or blown up in a terrorist act – which unfortunately has become a real day-to-day risk in most countries. Hence, it may be reasonable to consider the extent to which the risks associated with new activities increase our overall exposure to risk and the extent to which the risks associated with a new project relate to the overall risks experienced by society.

A study conducted for the East Japan Railway addressed a very specific question concerning perceived risks: to what extent should the consequences of catastrophic accidents be weighted more heavily in developing a risk management program for the railroad? This was a very practical question concerning the allocation of JR East's research budget and the amounts of money that should be spent on improving the protection at grade crossings in relation to the money spent on earthquake warning systems or installing collision avoidance systems for trains. The initial assessment of risks simply determined the expected fatalities per year from all kinds of accidents, which in effect weighted 100 fatalities in a train collision the same as 100 fatalities in 100 separate grade crossing accidents. However, the review of the literature cited above indicated that the public – you and me – really are more fearful of catastrophic accidents, especially those related to new technologies. The conclusion was that JR East should indeed spend proportionately more to avoid catastrophic accidents.

Indirect Consequences of an Accident or Incident

An accident or incident may result in a greater awareness of risks, which then leads to changes such as enforcing proper operations of the system (e.g. enforcing speed limits on highways), improving the infrastructure (e.g. retrofitting buildings and bridges so they will be better able to survive in an earthquake), or installing surveillance aimed at limiting deliberate attacks on the system (e.g. metal detectors at airports or procedures that limit public access to buildings). If these changes are indeed cost effective, then there will actually an improvement in the expected performance of the system relative to what was in place before the accident. Operating or equipment costs may indeed have gone up, but risks will have declined. What has changed is the understanding of the risks, which has resulted in rational decisions to reduce those risks. In this case, there would not be any indirect costs associated with the changes in operations, equipment, or regulation.

However, a very rational approach may be difficult in the turmoil following a catastrophic accident. People – system managers, government officials, passengers, and the public – all want to do something to prevent a similar accident. In what may become a highly emotional environment, there may be pressure to make changes that do seem to address the immediate problem, but that would not satisfy a cost-effectiveness test. Three types of improper reactions might be encountered:

- "Over-reaction": an accident or incident could result in changes that really are not related to or that go far beyond the specific causes of the accident or incident.
- "Irrelevant reaction": the proposed changes might in fact have little or no impact on risk.
- "Ineffective or counter-productive reaction": the proposed changes might increase risks or reduce the risk, but at only at a very high cost. Greater reductions in risk would have been gained by investing in other types of problems or other types of solutions.

Improper reaction leads to extra costs to the infrastructure operator and to society that would not have been incurred if there had not been a catastrophic accident. These costs are excessive, because they are above what could be justified using the standard probabilistic risk assessment approach. On the other hand, effective response to a disaster may give a company or an agency credibility with the public, thereby allowing a more rational approach to risk management – and to project design and evaluation.

Using Accident Statistics to Determine Causality

Whenever there is an accident (or an identifiable incident that could have led to an accident), there is an opportunity to collect data concerning the conditions at the time of the accident, the factors that led to the accident, and the consequences of the accident. This data can be used to learn more about the causes of accidents, which will be helpful in devising policies and technologies to reduce the number or the severity of accidents.

Three types of analysis are feasible:

- Analysis of data from thousands of accidents to understand the most important causal factors: this approach can be used for studying automobile accidents, because there are hundreds of thousands of accidents per year and there are well-defined policies for filing accident reports.
- Intensive analysis of each major accident in order to identify causality: this approach is used for accidents involving airlines, railways, and other transportation companies where any accident can be catastrophic.
- Statistical analysis of a representative sample of accidents: this approach can provide more accurate and more in-depth documentation of the factors related to the accident in order to understand the sequence of events and factors that lead to an accident.

If you are involved in an automobile accident, you are required to fill out an accident report that describes what happened, when it happened, and who was involved. If the accident is serious, then police will conduct their own investigation and file their report. The data from such reports is entered into a data base in order to allow statistical assessment of the causes of accidents. Data for hundreds or thousands of accidents can be used to determine how accident frequency varies by time of day, age of the driver, weather conditions, type and age of car, type and condition of road, and other factors. This data can be combined with other information about traffic volumes and the characteristics of licensed drivers to obtain estimates of accident rates, which could be measured as the number of accidents per million miles driven. This type of analysis shows that risks are higher for new drivers, very old drivers, drivers using cell phones, drunk drivers, drivers who speed, and drivers and passengers who do not wear seat belts.

Knowledge of these risks has led to public policies aimed at reducing the number or severity of accidents:

- Policies aimed at new drivers: requirements for driver training, restrictions on initial driver's license (limits related to time of day and number of passengers), and programs that increase public awareness of risks faced by new drivers.
- Policies aimed at very old drivers: eye tests for renewal of licenses; programs that increase public awareness of risks faced by old drivers (which may help older drivers realize that they need to limit or stop their driving); research to develop new technologies to enhance visibility or to reduce reaction time of older drivers.
- Policies aimed at reducing drunk driving: suspension or loss of license; public awareness programs; highly publicized extra enforcement on major holiday weekends.
- Policies aimed at reducing the severity of accidents: requiring seat belts and airbags; construction standards for cars; requiring helmets for motorcyclists.
- Policies aimed at improving the design of highways: limiting access to highways; minimizing curves; requiring wider lanes; eliminating dangerous intersections; improving sight distances by clearing vegetation; improving signage.

Detailed Analysis of a Few Major Accidents

It is also possible to conduct very detailed analysis of particular accidents or incidents in order to determine causality. Such investigations will be undertaken by company and government officials for any dam failure, any incident at a nuclear power plant, airline crashes, rail collisions and any other accident or incident that led to or could have led to a catastrophe. The results of the investigation may highlight problems with design, operations, maintenance, personnel training, weather or other factors. A single accident or incident may result in new regulations, new designs for equipment, inspection of an entire fleet of airplanes, or new training practices.

The standard accident report submitted does not provide extensive information about the condition of the vehicle or the driver and may have incomplete, confusing or inconsistent accounts of the accident. If the drivers and witnesses can be interviewed soon after the accident, it will be possible to learn more about not just what happened, but why it happened. If the driver works for a transportation company, then there may also be records that indicate how long the driver had been on duty and how often the driver had been driving in the past several days.

Performance-Based Technology Scanning

A project may be completed on time, within budget, and with no serious accidents, yet still be unsuccessful. The project may fail to operate as intended or, more likely, the actual demand for the project may be much less than expected. To minimize the financial risks associated with building projects that too few will use, it is essential to think about how people will respond to the project long before the design is finalized. The usefulness of the project and the ultimate demand for the project will be based, not upon the technical difficulties of its construction or the personalities of the leading characters promoting the project, but upon the actual changes in system performance made possible by the project. It is well to consider whether or not the proposed improvements in performance will really be sufficient to attract enough users and supporters who will be willing to pay enough to cover the costs of construction and operations. It is not necessary to know how to build a bridge in order to begin thinking about the potential benefits if the bridge were built. Likewise it is not necessary to know how to make pure drinking water more widely available in developing countries in order to estimate the health benefits that could be achieved.

The process of deciding what kinds of projects might be most useful is similar to the process of deciding what technologies might be most useful. If new technologies are to be adopted, whether the technologies involve electronics, bio-engineering or stronger materials, a series of projects will be needed to implement those technologies. Methodologies that have been used to determine which new technologies might be most useful turn out to be applicable in deciding what types of projects might be most useful. This section therefore introduces concepts broadly referred to as **technology scanning**.

The goal of technology scanning is to identify and evaluate new and emerging technologies that are potentially important to an industry, its competitors and its customers. **Performance-Based Technology Scanning (PBTS)** is a methodology for identifying areas where new technologies can have greatest performance benefits over the next 20-30 years in terms of reducing costs, increasing market share, and achieving higher profitability.[10] New technologies, however exciting, are germane only if a) the technologies lead to significant improvements in system performance (speed, safety, cost, ease of use, etc.) and b) those performance improvements are indeed important for users of a system. Effective technology scanning can help an industry identify new technological approaches, formulate a broader yet better focused R&D program, and improve its investment strategies. A superficial technology scanning program will readily identify exciting technologies, but it can be distracted and diverted into finding high-tech solutions for minor problems rather than seeking technological assistance in dealing with fundamental problems. A balanced technology scanning program should consider how technology can help meet customer needs and overcome fundamental operating constraints.

Table 2 shows the range of possible technology scanning activities as they might apply to companies or industries or agencies that manage or develop infrastructure systems. At the broadest level of technology scanning, there is a "General Search for Technologies". This search is of necessity somewhat unstructured, as it is not initially clear what new and emerging technologies will be available or what relevance they will have for the industry. This search should involve people with varied backgrounds and different working contexts, so that the search is truly broad. The three intermediate activities provide ways to narrow the technology scan from the "general search for technologies" to the "analysis of specific technologies".

"Technology Mapping" is the most general of these activities, since it predicts the effects of hypothetical technological changes in order to find the most important technological constraints on system performance. This activity, for example, might consider the relative advantages and disadvantages of increasing capacity, reducing costs by enabling more efficient operations, increasing the life of major system components by introducing more effective maintenance or inspection technologies, enhancing safety or security. Technology mapping begins with a base case that illustrates performance of a representative portion of the system. Next, high-level models predict performance for particular types of services as a function of technological capabilities, using inputs that capture the desired or anticipated results of deploying new technologies on each aspect of performance. Separate sets of models can be developed to predict

---

[10] Martland, C.D. "Performance-Based Technology Scanning Applied to Containerizable Freight Traffic," **Journal of the Transportation Research Forum,** Vol.57, No. 2, Spring 2003, pp. 119-134

the performance of typical types of systems at sufficient detail to provide realistic performance data and to capture the major competitive issues for different market segments.

Systems modeling is a more detailed activity. The objective here is to estimate the effects of particular technological improvements on system capabilities and performance. This is where very detailed engineering models could be useful.

Table 2 Performance-Based Technology Scanning

General Search for Technologies
  Conduct a very broad review of new and emerging technologies that might be beneficial to the system.
Technology Mapping
  Conduct structured investigations into the performance capabilities of the system and identify the points of leverage for technological developments related to cost, reliability, safety, or capacity of the system and any competing systems.
Systems Modeling
  Develop and maintain a set of models that can be used to evaluate technological improvements as they affect specific aspects of systems performance.
Customer Requirements Analysis
  Investigate the requirements of selected groups of customers or users and identify how new technologies might enable new ways of doing business; estimate the benefits to customers that will result from improvements in cost, speed, reliability, safety or capacity.
Analysis of Specific Technologies
  Examine specific technologies identified as having potential for improving system performance.

Customer requirements analysis can also be carried out at various levels of detail. There are several basic questions. How will a particular group of customers respond to potential changes in price, service, safety or capacity? What constraints, if any, limit the amount of services that will be purchased by these customers? How important are improvements in equipment design as opposed to improvements in trip times and reliability?

Finally, at the most detailed level of technology scanning, there is a need for analysis of specific technologies to demonstrate that a particular technology is indeed suited for the industry. Care is required in selecting technologies for this expensive stage of technology scanning.

The overall process is called "Performance-Based Technology Scanning (PBTS)". It includes consideration of the basic technologies, but also investigates how technologies translate first into better technological performance and then into better system performance in terms of the competitive market environment. The best technologies will relieve constraints that currently limit competitiveness. Using PBTS, it is not only possible to find new technologies, but also to identify gaps in an existing research program or to rearrange investment priorities in order to achieve more rapid implementation of the most effective technologies.

Summary

Project evaluation requires careful examination of both risks and uncertainty. Sensitivity analysis can help identify key variables, but more detailed methodologies are needed to understand risks and uncertainties related to engineering issues and the public's response to new technologies and new projects.

Simulation and Analytical Models

Models can be very useful in providing insight into the performance of a system, the likely effect of a project, or the amount of risk associated with a project. Models can be used to explore three key questions:

1. How will system capabilities or performance change if the project is implemented?
2. If system capabilities or performance change, what will happen to demand?
3. How will uncertainties affect the success of the project?

Models can be simple or complex, depending upon the stage of the analysis and the nature of the issues being investigated. Deterministic or probabilistic models can be developed, and even quite simple models can be helpful in determining the key factors affecting the success or failure of a project. By treating key factors such as project cost or demand as random variables, it is possible to see how the expected variations in these factors might interact and to estimate the likelihood that a project will or will not succeed.

Probabilistic Risk Assessment

Probabilistic risk assessment is a structured methodology for understanding risk, perceptions of risk, and how best to allocate resources for reducing risk. Any infrastructure-based system can experience accidents resulting from poor design, structural failure, bad weather, earthquakes, mismanagement, human error in operations, or other causes. The risk associated with any type of accident is defined to be the probability of an accident multiplied by the expected consequences of an accident. There are many possible consequences of an accident, including property damage, minor injuries, serious injuries and fatalities. If weights are applied to each type of consequence, it is possible to come up with a single measure of risk. The weights may be stated in monetary terms, in which case the weight can be interpreted as the value of a unit reduction in each type of risk.

Research into human behavior has shown that people are especially concerned with risks associated with unknown factors or catastrophic accidents. People apparently believe that more care should be taken to reduce the risks associated with potentially catastrophic accidents, such as a melt-down at a nuclear power plants or a chemical explosion than is necessary to be taken with respect to well known, but non-catastrophic accidents, such as occur on ski slopes or highways.

Risks can be summed for different types of accidents and different locations. Global risk of a system is the summation of the risks of all types of accidents over all locations within the system. Projects may increase some risks while reducing other types of risk. The increases or decreases in risk can be treated as costs or benefits of a project.

Public perceptions of risks are likely to differ from what engineers and risk experts calculate to be the risks. A highly publicized catastrophic accident – or even an apparently minor incident that could have been catastrophic - may cause public uproar and outrage. In the immediate aftermath of such an accident or incident, there could be extreme pressure for public action to ensure that such an accident "never happens again". Such a response could be an over-reaction that goes beyond the specific cause of the actual accident, an irrelevant reaction that has little or no impact on risk, or an ineffective reaction that may reduce risk, but only at an excessive cost. Infrastructure managers can reduce the likelihood of such improper responses by understanding the risks associated with their system, adopting and publicizing a risk management program, and responding quickly and effectively to any accidents or incidents that may occur.

Studies can be undertaken to increase the understanding of risks and thereby guide the selection of projects that will be most cost effective in reducing risks. For accidents that are common, such as highway accidents, it is possible to assemble a data base with basic information on every significant accident. The data base can then be used to support statistical analysis regarding accident causes or severity. For accidents that are rare and potentially catastrophic, such as those involving a power plant or a plane crash, every accident should be studied in great detail in order to determine whether there are previously unknown risks that need to be addressed. In a special study, it is possible to identify the critical event that led to each accident, the critical cause that precipitated the critical event, and the factors that were most likely to be associated with serious accidents. By understanding the causal chain that led to the accident and by isolating the most important associated factors, it is possible to formulate a strategy for reducing the risks of this type of accident, including recommendations for changes in design of new facilities or projects that will correct deficiencies in existing facilities.

Performance-Based Technology Scanning (PBTS)

Many projects involve the introduction of new technologies or the modification of an existing system in response to competitors' introduction of new technologies. Technology scanning refers to the search for new technologies that might be important to a particular industry. PBTS is a structured approach to technology scanning that focuses on the way the technology might affect system performance rather than on the details of the technology. In using this methodology, a technology can be represented as an option that has particular cost characteristics (e.g. investment cost, operating cost, lead time required for implementation) and performance impacts (e.g. reduction in operating cost, increase in capacity or improvement in level of service.) It is even possible to consider a range of hypothetical technologies with various cost and performance characteristics. Whether or not the technologies are useful will depend on a market analysis: will the changes in performance attract new demand? How will the new services be priced? How will competitors respond? Answering these question may require a great deal of analysis – but not necessarily any great detail concerning the technology. Methodologies developed for PBTS are readily useful for evaluating potential projects.

Case Study
An Engineering-Based Service Function for Bus Operations

This case study shows how a transit agency might develop an engineering-based service function that it could use to evaluate options for improving transit service. Using such a service function, the agency could determine how much service would improve for various operating and investment strategies. The cost effectiveness could be measured by comparing the improvements in service to the cost of the investments. By considering cost-effectiveness, the transit agency could determine whether it would be best to improve the existing services or to extend these services. They might conclude that the most cost-effective approach would be to increase bus frequency – but they might also eventually decide that they could not afford to do so.

Commuters in large cities usually have a choice of driving or taking transit. Public officials advocate greater use of transit as a way to reduce highway congestion, improve air quality, and reduce emissions of greenhouse gases. However, commuters will only take transit if they perceive the cost and service levels to be acceptable in comparison to driving. Transit agencies therefore need to understand how long it will take commuters to get from home to work if they decide to take the bus; if the transit agency can provide better service, then perhaps it can attract more riders. What is needed is an engineering-based service function that can predict trip time for a commuter based upon existing or potential transit schedules.

Any commuter's journey to work can be divided into individual segments. For example, one commuter's transit journey might include the following:

- Walk to bus stop (5 minutes)
- Wait for bus, which operates every 10 minutes (0-10 minutes)
- Ride bus 2 miles to subway station (5-10 minutes)
- Transfer from bus to subway platform (3 minutes)
- Wait for subway train, which operates every 5 minutes (0-5 minutes)
- Ride train 3 miles to destination, with five stops (12-15 minutes)
- Exit station and walk to destination (7 minutes)

The total trip could take as little as 36 minutes, if the connections are perfect and there are no delays on the bus or the subway. On the other hand, if the commuter just misses the bus and also just misses the train, and if there are delays for both the bus and the train, then the journey could take 55 minutes. The average trip is likely to be about 45 minutes.

After a couple of weeks taking this route, the commuter would know when to leave home in order to be on time for a meeting and when to leave home in order to experience the least delay. The commuter relies on experience and does not require an engineering-based service function.

The transit agency, however, has designed the bus routes, built the subway system, determined how many stops to make along the bus routes, and established schedule frequencies. They also have extensive experience in travel time along the bus routes and the time required at bus stops and at subway stations. They therefore can develop equations that could be used to predict service for any commuter under any set of operating conditions and any assumptions about future bus routes and subway extensions. They also know, from census data and surveys, where people live and where the jobs are. They can therefore select a representative sample of commuters and calculate the following for each of them:

- Access to the system
    - If distance from home to subway is less than 0.25 miles, assume the commuter will walk to the subway at an average speed of 3 miles per hour.

- If distance from home to subway is more than 0.25 miles, assume that the commuter will take a bus if there is a bus stop within 0.25 miles; the time to walk can be estimated assuming an average speed of 3 miles per hour.
- If there is neither a subway station nor a bus stop within 0.25 miles, assume that the commuter will drive to work.
- Wait for bus (assume to be between 0 and the average time between buses during rush hour)
- Ride bus to subway station (estimate time based upon observed travel times, scheduled frequency of stops, and time required for stops along route)
- Transfer from bus to subway platform (estimate based upon distance from bus to platform, expected congestion at stairways or escalators, and time required to purchase ticket)
- Wait for subway train (assume to be between 0 and the average time between trains during rush hour)
- Ride train to destination (estimate time based upon distance, speed limits, train acceleration and braking capabilities, number of stops, and time required for loading and unloading at each station)
- Exit station and walk to destination (estimate based upon distance and walking speed)
- Sum all of the times to determine the minimum, expected, and maximum travel times for each user.

This is an engineering-based service function because it provides a way to predict trip times and reliability as functions of engineering parameters, operating strategies, and network design. This process could be used to estimate the changes in service, for particular groups of people, that would result from increasing the frequency of bus service, adding bus routes, creating an express-bus lane, or expanding the subway system.

Figure 1 Reliable, frequent bus service in England's Lake District
Visitors can leave their cars at home, take the train to Windermere, and connect to buses that offer frequent, reliable service throughout the entire Lake District. The views of the mountains and lakes are more important than speed.

Case Study
Capacity of a Highway Intersection

This case study explores the capacity of a highway intersection in order to illustrate the differences among maximum capacity, operating capacity and sustainable capacity.

Consider a typical intersection of two arterial streets in an urban area, one of which carries eastbound and westbound traffic, while the other street carries northbound and southbound traffic. To simplify the situation, assume that a) all of the vehicles are automobiles, b) none of the vehicles are making any turns, and c) there are no pedestrians. The traffic signal has a cycle time of 120 seconds, and it provides equal time (60 seconds each) for the east/west and the north/south flows. When the signal turns green, it takes 2 seconds for the first car to move through the intersection; if there is a queue of cars, then additional cars can pass through every 2 seconds. Thus, the maximum number of cars in one lane that can move through the intersection in each two-minute cycle is:

(60 seconds green/cycle)/(2 seconds/car) = 30 cars/cycle

If both roads are two-lane roads, then there would be a total of 4 lanes that approach the intersection and there could be as many as 4 lanes multiplied by 30 vehicles per lane or 120 vehicles moving through the intersection during each two-minute cycle, and there could on average be one vehicle per second moving through the intersection.

Is the capacity of this intersection therefore one vehicle per second or 60 vehicles per minute? If so, is this the same as saying that the capacity is 60 vehicles per minute or 3600 vehicles per hour or 24 hours/day (3600 vehicles/hr) = 86,400 per day?

Given the assumptions, these may appear to be reasonable statements. However, we probably should be somewhat suspicious of the nice round numbers used in the above analysis. If you really want to know the capacity of the intersection, you could stand on the corner and count cars. If you had a stopwatch and a clipboard, you could record the actual number of cars moving through on each cycle and you perhaps would find that sometimes the first car takes 4-5 seconds (until the driver behind honks), while subsequent cars take a little more than two seconds. With better information about how cars move through the intersection you would be able to say something like this: the maximum number of cars passing through the intersection at rush hour during one week in October was 56; the average number of cars passing through the intersection was 52. Moreover, you noted that there was never enough time for all of the waiting cars to get through in one cycle, so that these observations were directly related to the capacity of the interchange. Armed with this knowledge, you might conclude that capacity was somewhat less than 60 cars per minute. And you might wonder whether the capacity should be stated as 60 (which is what was calculated using the textbook approach described above), 56 (the maximum that you yourself saw) or 52 (the average that you observed)?

Before answering this question, let's think about how this measure of capacity might be used. If we are traffic engineers or planners, we likely would be comparing the capacity of the intersection to the volume of traffic that is expected to be trying to get through the intersection; if capacity is insufficient, we would then be considering whether it is worthwhile to expand capacity or to reduce demand (e.g. by establishing tolls or by promoting the use of transit). If we are commuters, we are likely worried about excessive delays if there is insufficient capacity. Engineers, planners, and commuters would all be thinking about the extent of the delays, and they would all be concerned with what happens during the morning and afternoon rush hours. Thus, they would be thinking about cars per hour rather than cars/minute. While all might agree that the theoretical capacity is somewhere between 56 and 60, they would all prefer to use no more than the observed average of 52 in their comparison of capacity to demand.

Is there any reason to consider something less than 52 cars/minute as the estimate of capacity? Yes, indeed. Any experienced commuter would have some questions about the study that you completed:

- Were there any accidents during that week?
- What was the weather like (were there any heavy rains or snowstorms)?
- Were there any emergency vehicles trying to move through the intersection?
- Was there any maintenance in or near the intersection?
- Was the intersection ever gridlocked (i.e. did cars ever enter the intersection without being able to continue all the way through, thereby blocking the intersection when the light turned green for the other traffic)?

Commuters ask these questions because they know that accidents, bad weather, emergency vehicles, and aggressive drivers all disrupt normal flows, not necessarily every day but often enough to be a problem. A more extensive study that included winter conditions and periods of maintenance might show that the average flow through the intersection was actually only 48 cars per minute during rush hour.

So is the capacity of the interchange 48 cars per minute? Well, the commuters and the planners would ask another question: how bad are the delays at this intersection? It could be that this is a notorious intersection where cars back up for a half-mile or more and it often takes 10-15 minutes to get through. Anyone driving that route on a regular basis would say that such delays are unacceptable, and planners would likely agree that the intersection was experiencing traffic volumes that were higher than its capacity. This may seem to suggest that the capacity is less than 48 cars per minute, but actually it doesn't. The capacity of the intersection is in fact about 48 cars per minute, but traffic volumes during rush hour are greater than this; the long queues build up because cars arrive faster than cars can get through the intersection. When an incident happens, whether related to an accident or to a snow-storm or to anything else, fewer cars move through the intersection and queues build up to an unacceptable level.

A traffic engineer might then ask the following question: if the average throughput of the intersection is 48 cars/minute during rush hour, what is the maximum traffic volume for which the performance of the intersection is acceptable? This question could be phrased in terms of the structure of the street network: what is the maximum traffic volume for which the queues at this intersection will interfere with the performance of neighboring intersections no more than once or twice a month? This question could be answered using algebraic models or by simulating traffic conditions. If a delay of a few minutes is acceptable, then it will certainly be possible for more than 48 cars/minute to be arriving during rush hour even though only 48 cars/minute will get through the intersection.

The discussion thus far has only considered rush hour, and the estimates of capacity all related to what happens during rush hour. In designing systems, it is essential to consider capacity during the peak period, but estimates of demand may be phrased in terms of daily, weekly, monthly or annual usage. It is therefore necessary to be able to translate peak period capacity into capacity for these other periods. This can be done by considering the pattern of demand over longer periods. For highways, the following relationships can be estimated based upon observations of traffic volumes:

- Average vehicles per day on week days over an entire year
- % of traffic on weekdays
- % of traffic during peak periods of the day

For example, if the capacity of an intersection is 50 vehicles/minute during rush hour, then the capacity for the four peak hours will be 12,000 vehicles during four peak hours. If the rush hour periods account for 40% of weekday traffic, then the intersection can be considered to have capacity for 12,000/0.40 = 30,000 vehicles per weekday. If weekdays have 80% of the average weekly traffic, then the capacity is 30,000 vehicles per weekday x 5 weekdays/week / 0.80 = 187,500 vehicles per week or 9.375 million per year, assuming that the current traffic patterns remain the same. This statement can be interpreted as follows: if traffic along these two roads rises to 9.4 million per year, then this particular intersection is likely to experience unacceptable delays on a regular basis. The intersection might actually handle more traffic, but the delays would increase and the congested rush hour periods would be longer. In other words, this intersection is expected to act as a bottleneck once traffic along these two roads rises to 9.4 million vehicles per year.

The capacity of the network will be harder to define, because there will be many possible bottlenecks. There may be severe problems at other points in the network long before our intersection experiences significant delays. If it is easy to add capacity, then it may be possible to add capacity as demand grows, thereby avoiding the development of bottlenecks.

If there are many potential bottlenecks within the network, then fixing one of them will not necessarily affect system performance. Increasing the capacity of one intersection by building an overpass may simply shift the delays to the next intersection.

Case Study
Canal Projects in the Early 19th Century

Before the invention of railroads, canals offered the most cost-effective alternative to shipping freight by horse-drawn carts along poorly maintained gravel roads. This case develops engineering-based cost and service functions for both modes of transportation in order to examine whether or not an investment in canals was likely to be attractive to potential users and therefore profitable for potential investors.

Introduction

Canals were among the first major civil engineering projects in the United States, as in many other countries. Until railroads were invented, land transportation was cumbersome, slow, and expensive. Water transportation – when available - was much cheaper, more reliable, and provided the only means of handling large volumes of freight, which of course was why cities grew up at the best harbors and along the major navigable rivers. The first canals simply bypassed rapids so as to avoid costly transshipment of goods. Later canals linked major cities to their hinterlands. The early canals were designed with a tow path on either side so that a horse or a team of horses could pull the canal boats along the canal (Figure 1). The average speed would be only 2-3 miles per hour, and the distance traveled per day would only be 20-30 miles. More ambitious projects, such as the Erie Canal, sought to open up western regions and thereby promote development (not to mention the importance of the port city whose citizens promoted the project).

Figure 1 Cross-section of a typical 19th century canal.
The canal would be barely wide and deep enough to allow two small boats to pass. The boats would be towed by a horse that walked along the tow path next to the canal.

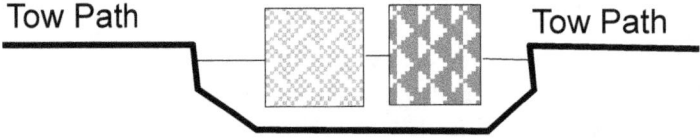

Canals must be close to level to allow safe, easy movement in both directions. To create a level route in uneven topography, it is necessary to construct locks. A lock is a chamber with two sets of gates. The water level is higher on one side of the lock than it is on the other side. By opening one gate at a time, it is possible to adjust the water level within the chamber to match either the high side or the low side. The depth of the lock must be sufficient to accommodate canal boats when the water is low; the height of the gates above the low water level limits the extent of the lift that can be achieved by one lock. Several locks can be operated next to each other or in close proximity to each other if the terrain requires more lift than can be achieved with a single lock. The time required for a move through a lock includes the time to position the canal boat (or boats) in the lock, the time required to raise or lower the water, and the time required for the boat(s) to move out of the lock.

Figure 2 Remnants of the C&O Canal can still be seen in Georgetown in the District of Colombia. Note the long narrow shape of the boat, which is sized to squeeze into the locks, such as the one at the back of this picture.

The width and depth of a canal and the size of the locks determine the size of the boats that can use the canal. Figure 2 shows a lock on the Chesapeake and Ohio Canal in the Georgetown neighborhood of Washington D.C. Note that the lock is much narrower than the canal, and it has a lift of less than 10 feet. There is only one channel in this location, but if more capacity were required a second lock could have been built right next to this one.

The deeper and wider the canal, the more material that must be excavated and the more expensive the project (Figure 3). The larger the locks, the more expensive they become and the more water that is required to operate the system. Hence, there are fundamental design issues concerning the size of the canal and the type of boats that will be accommodated. The canal must either be wide enough for two boats or provide periodic basins where opposing boats can pass. It must also be deep enough to provide the required draft for the largest boats that are allowed to use the canal. Some of the early canals could only handle small boats with a capacity on the order of 15 tons; these boats required a draft of only twelve inches when loaded. Larger canals could handle larger boats, e.g. boats that could carry 75 tons along canals providing more than four feet of draft. As suggested by Figure 3, increasing the width and depth of a canal can require massive excavations in hilly regions.

Figure 3 Topography has a large effect on the costs of constructing a canal.

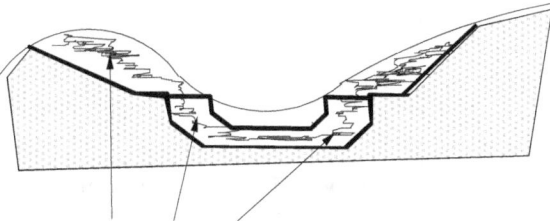

Doubling the width and depth of the canal can lead to major increases in excavation

The number of locks required is a function of the route and the size of the locks. The topography of the route determines the minimum lift that will be required, which is the difference in elevation between the beginning and end of the canal. If a canal is constructed to connect two river basins, then the canal will have to either cut through the intervening hills or move up, across, and down the other side in a series of locks (Figure 4). Excavating a level route reduces the number of locks that will be required, which will also reduce the time required to move along the canal. However, it may be infeasible or extremely expensive to create a level route.

Figure 4 Locks enable canal boats to move up and over hills

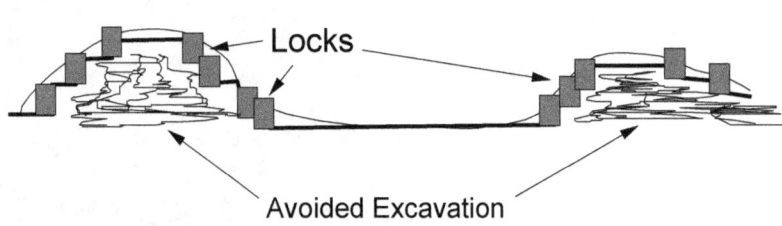

Each time a boat moves up stream through a lock, the lock fills with water; when a boat moves downstream, the lock is emptied. If a canal is to move from one watershed to another, there must be a reliable water supply in order to support the functioning of the locks (Figure 5).

Figure 5 A water supply is needed above the highest point of the canal in order to supply the water necessary to operate the locks.

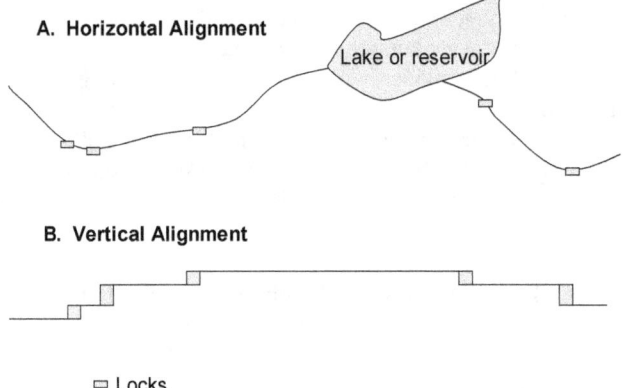

Service and Capacity of a Canal

Customers who might use the canal to move freight would be concerned with two aspects of service: the time that it would take to move along the canal and the maximum size for a canal boat. First let's consider the time that it would take to complete a 120-mile trip along a nineteenth century canal. We can assume that boats could tie up for the night at either end of the canal or at frequent locations along the length of the canal (as shown in the photo of Regent's Canal in Figure 6). At two miles per hour, a 120-mile trip would require 60 hours of travel time, which would be six ten-hour days; at three miles per hour, the same trip would require only 40 hours of travel, or four ten-hour days. Thus a reasonable estimate of the travel time could be stated as "five days plus or minus a day" or "less than a week". This may strike you as a fairly imprecise estimate, but before worrying too much about how to refine it, think about what else goes into the travel time. First of all, there are likely to be locks located every few miles along the canal, and it will take an hour or two to get through each lock, perhaps much longer if the canal is very busy and there are queues of boats waiting to get through the busiest locks. If the 120-mile trip has 40 locks, then the time spent in the locks could well be 40-80 hours, which is as long as and more variable than the travel time along the canal. There is also the possibility that bad weather, high or low water, lock maintenance, or other problems will limit or prohibit travel along the canal. Thus, the estimate that it will take about 50 hours to travel along the canal between locks is probably the most reliable portion of the overall estimate of travel time. Adding in locks and considering the possibility of major weather-related delays, the trip would likely be estimated as a journey of "two to three weeks".

Figure 6 Regent's Canal, London
England's canals remain busy, but the old canal boats have been converted to house boats suitable form mobile homes or for weekend get-aways. Dozens of boats may tie up for the night at a convenient and picturesque spot such as this.

The capacity of a canal (maximum throughput measured in tons of cargo) can be estimated for various time periods and conditions:

- A peak day with twelve hours of operation
- A peak summer month with twelve hours operation, seven days/week
- A year, with operations ceasing during the winter and during major storms.

The canal's capacity will be a function of the characteristics of the canal, the boats, and the operating characteristics. The maximum capacity of the lock is the inverse of the function of the cycle time for the lock. If the cycle time is a half hour per boat, then the maximum capacity is two boats per hour. The operating capacity, expressed in boats per day, will be limited by the hours of operation. If the lock operates twelve hours per day, then the operating capacity would be 24 boats per day. The sustainable monthly capacity would be further reduced to allow time for maintenance, to allow for periods of slower or interrupted operations during bad weather, and to allow for smooth functioning of the canal despite the normal variations in traffic and the routine delays that might occur. For example, the lock might be closed for one day per month for routine maintenance, heavy rains or winds might reduce capacity by 50% for several days per month, and miscellaneous delays might amount to an hour or so each day. This would reduce sustainable monthly capacity as follows:

- Days per month: 30
- Days required for maintenance: 1
- Expected days of operating at 50% capacity: 3-4 days, which is equivalent to 1-2 days of lost operation
- Miscellaneous delays: about 1 hour per day or 30 hours per month, which is equivalent to about 4 days per month

- Net days available for normal operation per month: 30 − 1 − 2 − 4 = 23 days

Thus the sustainable capacity would be no more than 75% of the operating capacity. A further reduction in sustainable capacity might be necessary based upon traffic patterns and service requirements. For example, traffic volumes might be much higher during the middle of the day and much lower on weekends, and users of the canal might expect at most modest delays during normal operations (i.e. delays related to maintenance or bad weather may be acceptable, but extensive delays related to congestion may be viewed as unacceptable). These considerations would reduce sustainable monthly capacity to less than 70% of operating capacity:

- Weekend days per month: 8-10 days, with perhaps traffic at 50% of normal volume, which is equivalent to 4-5 days of normal operation or about 15% of monthly capacity.
- Peak patterns of traffic: the lock must be functioning reasonably well during these peak periods, so there will be unused capacity the rest of the day. If peak period totals only 6 hours per day, then there will be idle capacity the rest of the day. If the off-peak volume is 50% of the peak volume, then the maximum daily capacity under normal operations and normal service levels will be only 75% of the operating capacity (100% for 6 hours and 50% for 6 hours produces an average utilization of 75% for the entire 12 hours).

If you consider these two factors along together, there will be a further reduction in capacity:

- Operating Capacity: 24 boats per day through the lock
- Sustainable capacity, without considering traffic patterns: 70% of operating capacity
- Adjustment for traffic patterns:
    - Weekends: 15% reduction
    - Weekday peaks: 25% reduction
- Sustainable capacity, taking into account traffic patterns: if these factors are considered to be independent, the sustainable capacity will drop to less than half of the operating capacity: $(0.70)(0.85)(0.75) = 45\%$ of operating capacity, or 11 boats per day.

However, there is certainly some overlap among these various factors. Storms may occur on weekends, and if they do, then they will not disrupt operations as much as if they occur on weekdays. Providing capacity for acceptable service during peak periods may also make it easier to schedule routine maintenance during slow periods of the day or the weekends. Miscellaneous delays that occur during off-peak times will not seriously affect capacity. The estimate of sustainable capacity is therefore perhaps 50-60% of the operating capacity or 12-15 boats per day or 360 to 450 boats per month.

Is this an acceptable estimate? Do we really believe all of these assumptions? Wasn't that just hand waving and magical thinking when the sustainable capacity was increased from the calculated 45% to a rather broad range of 50-60% of the operating capacity? The answer to all of these questions may be "maybe"! We could study canals in much greater detail in order to refine the assumptions, and we could develop simulation models to look much more closely at the effects of traffic patterns on service levels. However, rather than spending a lot of effort trying to answer these questions about methodology, let's look at several more aspects of capacity, namely the loads carried on each canal boat, the ability to operate throughout the year and the ability to adjust operations as needed to keep up with demand. Perhaps the estimate of sustainable capacity is acceptable as it is.

First of all, the size of the average load is very important. The maximum capacity measured in tons/day could be based upon the size of the largest boat that could use the canal. So, if the canal could be used by boats that carry 15 tons, then the maximum capacity would be 24 boats/day x 15 tons/boat = 360 tons/day. If the largest boat could carry 50 tons/day, then the maximum capacity would be 1200 tons/day. In either case, the average load – and therefore the operational capacity - would be considerably less than the maximum load, especially if some of the boats only carry loads in one direction. If the average load were estimated to be 11-12 tons/boat, that would be a reduction of 20-25% in capacity measured in terms of tons/day – and another reason for being somewhat imprecise in our measure of capacity.

Weather is an even larger factor. Canals are unusable when the water levels are too high or too low or when they are frozen over. Depending upon the climate, periods of inoperability could last for months or half the year, and these periods could vary greatly from one year to the next. If weather shuts down the canal for three months, then annual capacity would be reduced by 25%; if weather shuts down the canal for four months, then annual capacity would be reduced by 33%. There could easily be a 5-10% variation in annual capacity depending upon the weather.

Finally, the users and operator perhaps have considerable options regarding operating and pricing policies. All of the above estimates assumed operations of twelve hours per day. Is this a credible constraint? Would it not be possible to add some extra hours of operation during the peak season? And couldn't some of the off-peak capacity be utilized by reducing lock fees for these periods in order to promote somewhat different usage patterns?

Hold on! These are too many questions for this stage of the analysis! What can we conclude with some certainty? We think that the locks on the canal can probably handle close to 24 boats/day when things are going well, but that they may not be able to handle even half that much on a sustainable basis because of a variety of potential problems. We also think that the canal will be able to operate for about eight months of the year. If we take ten boats per day as the sustainable capacity, that would be 150 tons/day and 36,000 tons in eight months. We know this is a rather fuzzy number, but perhaps it will be sufficient. Now let's turn to costs and competition. Will there be a market for the canal?

Canal Competitiveness

When canals competed with horse-drawn wagons, they had a marked advantage in cost, as illustrated by the immediate success of the Middlesex Canal when it opened for operations between New Hampshire and Boston in 1803:

> *"The advantages of canal travel over wagon transport were obvious at once. One horse, for example, could easily draw 25 tons of coal on the canal. On land the same horse could pull only 1 ton. One team of oxen could pull 100 tons, an amount that would take eighty teams on land. In the first eight months of the canal's operation, 9,405 tons were carried at a cost of $13,371. The cost for such a shipment by land would have been $53,484."*
>
> Daniel L. Schodek, "Landmarks in American Civil Engineering", MIT Press, p. 12

These estimates of operating cost for this 27-mile canal can easily be converted to the cost/ton-mile of transporting freight by canal boat or by wagon. Expressing cost as the cost per ton-mile is useful because that allows a normalized comparison among different modes of transportation, different lengths of haul, and different time periods. Assuming that all of the 9,405 tons were transported the entire length of the canal, the cost per ton-mile of using the canal was $13,371/(27 miles x 9,405 tons) = $0.056 per ton-mile. If the distance by road was also 27 miles, then the cost of using a wagon was four time as large: $53,484/(27 miles x 9,405 tons) = $0.21 per ton-mile. These numbers are very interesting because they are considerably larger than the costs of transporting freight along the inland waterways and highways in the 21st century! Transporting coal on the inland waterways or on railroads now costs less than $0.02 per ton-mile, while the cost of truck transportation is well under $0.10 per ton-mile (and a penny in 1803 was worth a whole lot more than a penny is now worth 200 years later!)

Engineering-Based Cost Model for a Canal

Knowing the cost/ton-mile is interesting historically, and it is a useful indicator that customers or managers might use. Being able to estimate the cost/ton-mile as a function of design factors and operating conditions is essential to planning and evaluating a transportation project. It is mildly interesting to know that the Middlesex Canal cut costs by 75% relative to horse-drawn wagons, but a canal designer and his financial supporters will want to understand the potential costs and benefits related to constructing a specific canal that would attract traffic from wagons. Given assumptions about unit costs and productivity, it is possible to create an engineering-based cost model for a canal. Let's assume that the operating costs and productivity parameters for a canal are based upon typical values for the early 19th century:

- Cost for the two-person boat crew ($1/day each, for ten working hours)
- Cost for the teamster and the horse ($1/day each or $2/day total)
- Cost for the boat ($2/day for a boat with a capacity of 15-tons)
- Cost for lock operations ($2/day for an operator and routine maintenance)
- Cost for canal and embankment maintenance ($40/year per mile)
- Average speed (three miles per hour along the tow path)
- Average time per lock (12 minutes for a 15-ton boat)
- Annual operations: 225 days per year
- Annual tonnage along the canal: 10,000 tons

Using these assumptions, we can estimate the variable costs of operation for a canal of any length and with any number of locks. For example, let's calculate the variable cost/ton-mile for a trip by a 15-ton canal boat along a 30-mile canal that has 10 locks. First we need to know how long the trip will take:

- Travel time along the canal: 30 miles/3 mph = 10 hours
- Time in the locks: 10 locks x 0.2 hours per lock = 12 hours

The boat operates 10 hours per day, so the trip will take 1.2 days. Perhaps the crew finishes unloading in morning at one end of the canal and hopes to reach the other end in time to start unloading the following afternoon. We will assume that the cost of the trip will include 1.2 days for the crew, the boat, the teamster and the horse:

- Boat Crew: 2 people x $1/day x 1.2 days = $2.40
- Teamster and horse: one team x $2/day x 1.2 days = $2.40
- Boat cost: $2 per day x 1.2 days = $2.40
- Total variable cost = $7.20
- Total ton-miles = 15 tons x 30 miles = 450 ton-miles
- Variable cost per ton-mile = $7.20/450 = $0.016/ton-mile
- Variable cost per ton = $7.20/15 = $0.48/ton

The fixed costs include the cost for the lock operators (who are assumed to be available whether or not there is any traffic) and the cost for canal maintenance.

- Lock operations: ($2/lock per day) x (10 locks) x (225 days per year) = $4,500 per year
- Maintenance: $40/mile/year x 30 miles = $1,200 per year
- Total fixed cost: $5,700 per year
- Total fixed cost per ton: ($5,700/year)/(10,000 tons/year) = $0.57/ton
- Total fixed cost per 15-ton load: $8.55

The total, fully allocated cost for the trip will be the sum of the variable cost and the allocated portion of the fixed cost: $7.20 + $8.55 = $15.75. The total, fully allocated cost per ton-mile will be $15.75 per trip/450 ton-miles/trip = $0.035. This cost is well below the estimated cost of $0.21 per ton-mile of using a horse and wagon, so it is likely that the canal could attract traffic and earn profits for the investors.

If these cost and capacity calculations used in this example were incorporated within a spreadsheet, it would be possible to explore how costs and capacity would vary with differing assumptions concerning the structure of the canal, the size of the canal boats, the operating parameters, and the unit costs. Such a model could also be used to see how costs/ton-mile would vary with the annual tonnage on the canal and the size of the boats.

Evaluating the Canal Project

From the designer's perspective, critical questions would concern the size of the canal: should they build it only to accommodate 15-ton boats, or should they build it to handle larger boats? To handle larger boats, the canal would have to be a little wider and deeper, and the locks would have to be larger, so the costs of construction would rise. On the other hand, with larger boats, the variable costs/ton-mile would be expected to decline, since the same crew could handle a larger boat.

From the owner's perspective (or the perspective of the banks and investors who were providing the funds for the project), the key question would be whether or not they could charge tolls sufficiently large to cover their fixed costs of $5,700 per year plus sufficient profit to justify their investment. To figure out how much they would need to charge for a toll, we need an estimate of what it would cost to build the canal. Canal costs were on the order of $20,000 per mile in the early 19th century, so that a 30-mile canal would have cost about $600,000 to construct. This money would have to be raised from investors who would expect (or at least hope) to receive a substantial annual dividend. A 10% dividend would amount to $60,000 per year, so the total amount of money raised by the toll would have to be $65,700 per year. If this were to be raised by a toll based upon tonnage, the average toll would have to be $6.57 for the expected 10,000 tons/year.

Is this a realistic toll? To answer this, we need to consider the perspective of the user. By using the canal rather than a horse and wagon, the user transports freight at a variable cost of $0.48 per ton. The toll would raise this cost to more than $7/ton or $0.25/ton-mile for the 28-mile canal– which is more than the cost of using a horse and wagon, which was estimated above to be $0.21/ton-mile! With such a large toll, the canal would have difficulty attracting any traffic at all.

Upon hearing this sad news, the investors would have several options. They could cancel the project as unprofitable. They could settle for a smaller return on their investment; cutting the toll to $3.57 per ton would provide a 5% return, and it would keep the cost per ton-mile well below the cost of the competition. They might also conduct a more careful study of demand, including an assessment regarding the potential for growth in canal traffic over a 10- to 20-year horizon. They could also consider building the canal for larger boats, a strategy that would further reduce operating costs for users, possibly for a fairly modest increase in construction costs.

Notice how the degree of precision has softened as we have progressed through this example. We started with some concern over the many assumptions that we were making concerning capacity and operating costs. By the time we got to the end, however, we discovered that the largest cost by far would be the return on investment that would be required to attract investors. Given the cost of the canal, the required toll would be an order of magnitude larger than the users' operating costs, and the major problem with the project appears to be that there is not enough traffic to justify the investment.

To reach this conclusion, it doesn't matter much if at all whether the canal boats move along at 2 or at 3 mph, and it doesn't matter whether the locks take 12 minutes or 15 minutes. It also doesn't matter whether the capacity of the canal if 50% more or less than our preliminary estimate of 36,000 tons per year. Why? Because our projected demand is barely 25% of that amount; adequate capacity seems to be assured. The lack of precise estimates of operating costs and capacity doesn't matter nearly as much as what we have discovered to be much more important considerations:

- How much traffic will use the canal – when it first opens and over the next ten to twenty years? Perhaps the investors can defer dividends for a few years in order to secure very attractive profits in the future.
- What will it really cost to build the canal and how great a return on their investment will investors require?
- Should the canal be redesigned to handle larger boats that might attract more traffic?

Examples of Canals

China constructed its Grand Canal more than 1300 years ago[1]. Linking Beijing with the country's major river systems and ultimately the coast, this canal provided a means of transporting a steady supply of grain from the south to the north of the country. During the 7th century, 300,000 tons of grain were transported per year along this route. The canal was an enormous undertaking: 5.5 million laborers worked six years on one 1,500-mile stretch of the canal (20 person-years per mile).

England's industrial revolution was given a strong push when canals were constructed that provided cheap transportation for coal, agricultural products, and everything else[2]. The Bridgwater Canal, built in 1761 to link Manchester with coal mines, halved the price of coal in Manchester and helped Manchester become England's leading industrial center. By the 1840s, the country had a network of 5,000 miles of canals and navigable rivers, and nearly every city or town was within 15 miles of a canal. As the country prospered, canals were built to be straighter, wider, and deeper; aqueducts were constructed to allow canals to cross rivers.

The Potowmack Canal was the first extensive system of river navigation in the US[3]. Championed by none other than George Washington, the canal was designed to open up the area west of the Appalachian Mountains by providing a water roué to the Potomac River (which flows toward what is now Washington D.C.). The canal allowed boats with a 16-20 ton payload to make the 185-mile trip in three days at half the cost of transporting the same freight by horse and wagon. The canal ran into problems because of recessions, lack of skilled workers, and bad weather. The route was navigable only for three months of the year, the canal tended to fill up with sediments, and the wooden locks decayed. The canal did help spur investment in and development of the western region, but it was a financial failure. After an investment of $750,000 between 1785 and 1802, the canal company was $175,000 in debt by 1816.

The Middlesex Canal was a similar project that was aimed at providing a better link between Boston, Massachusetts and the farms and forests of New Hampshire[4]. The 27-mile canal required the construction of 50 bridges, 8 aqueducts, and 27 locks. The investment of $528,000 ($20,000 per mile) was, for the time, an enormous amount equal to 3% of the assessed value of all property in Boston. The canal suffered because the freight – what little there was - was mostly southbound. Despite the small volume of freight, there were political disputes, as people in New Hampshire did not appreciate a company chartered by Massachusetts diverting freight from New Hampshire's major port.

The Erie Canal was the most ambitious canal project undertaken in the US during the 19th century[5]. The project was first proposed in 1724, and it was widely discussed for nearly 100 years as a means of linking New York City and the Hudson River with the Great Lakes at a point to upstream of Niagara Falls. Because of the geography of the eastern US, the Erie Canal route was the easiest way to get from Atlantic ports across the Appalachian Mountains. Thomas Jefferson called the Erie Canal "a splendid project – for the 20th century!"

Gaining support for the construction of the canal required a major political effort, and not only because of the difficulty in financing such a large project. There was uncertainty about the route; an inland route would be expensive, but it would avoid exposing trade to attacks from Canada if the route used Lake Ontario. Merchants who used ground transport were against the project, because it would expose them to competition. There was also a lack of skilled engineers for carrying out the project – and the project therefore led to the creation of civil engineering schools at Rensselaer Polytechnic Institute and Union College.

DeWitt Clinton was a member of the commission formed in 1810 to consider the construction of the canal. As a former mayor of New York City, a US senator, and eventually as governor of New York, Clinton was the foremost champion of the project, and it was finally approved by the state legislature. The 363-mile canal, with 83 locks and

---

[1] E.L.Newhouse, Ed., "The Builders", National Geographic Society, Washington DC, 1992, p. 29
[2] E.L.Newhouse, Ed., "The Builders", National Geographic Society, Washington DC, 1992, p. 30
[3] Schodek, "Landmarks in American Civil Engineering", MIT Press, Cambridge, MA pp. 3-6
[4] Schodek, op. cit., pp. 5-12
[5] Schodek, op. cit., pp. 13-19

18 major aqueducts between Albany and Buffalo, was constructed between 1817 and 1825 at a cost of $8 million plus the loss of 1000 lives from malaria and pneumonia.

In order to limit costs of construction, the canal was built just wide enough (40 feet) and deep enough (four feet) to handle medium-sized boats. When the canal opened, demand and revenue exceeded all expectations, and it was possible to finance projects that increased the canal to a width of 70 feet and a depth of seven feet.

The long-term impact of the canal was immense. According to Schodek, opening up Lake Erie was the "decisive impetus" for commerce in the eastern US to move east-west rather than north-south. Rochester and Buffalo became boom towns, and population in New York increased all along the route of the canal. The success of the Erie Canal sparked development of a system of canals in Ohio, as population and economic development in the country pushed further westward.

Epilogue: Canals vs. Railroads

Table 1 summarizes the costs and operating characteristics for the main transport options in the early 19th century. Turnpikes provided a way to achieve substantial improvements over rough roads, and they could easily be financed by tolls. Canals cost two to four times as much to construct, but they cut freight transport costs to a third of the costs of using turnpikes. Canals enjoyed only a brief period of supremacy in the United States, as they were restricted by topography and their service was much slower that what was possible once railroads were introduced. Railroads were similar to canals in terms of construction cost and operating cost, but they were much faster and therefore much more attractive for passengers and for most kinds of freight.

Table 1 Comparison of transportation Costs, first half of the 19th century

Rough Road $1-2,000/mile to construct	1 ton/wagon 12 miles/day 12 tm/day/vehicle	$0.20 to $0.40/tm for freight rates
Turnpike $5-10,000/mile	1.5 tons/wagon 18 miles/day 27 tm/d/v	$0.15 to $0.20/tm
Canal >$20,000/mile	10-100 tons/boat 20-30 miles/day 200-3000 tm/d/v	$0.05/tm
Railroad $15-50,000/mile	500 tons/train 200 miles/day 100,000 tm/d/v	<$0.05/tm

Lessons for Other Infrastructure Projects

This brief review of a few of the major canal projects in the US provides some useful lessons regarding projects:

- Ideas and concepts for major infrastructure projects may abound long before the means to build the infrastructure are available.
- Important public figures may become champions for particular projects.
- Major projects can, like the Erie Canal, be decisive in directing development and population growth, but it is also possible to spend major resources on projects like the Potowmack and Middlesex Canals that have only modest potential.
- Changes in technology can kill projects (railroads quickly put these canals out of business by the mid-19th century).

- Financing is a major concern.

The discussion of canals also offers some insight to the different perspectives of the various participants in evaluating projects. If potential users' costs are lower, they will use the facility. In deciding whether to use a new canal, potential users had to ask whether it would be possible to lower freight costs by using canals rather than horse and wagon or by using large canal boats rather than smaller boats. Potential users therefore would compare their costs for equipment and operations for their current and newly available options.

Owners or entrepreneurs have a different question: should they build the facility? They have to compare annual revenues to annual costs, taking into account the costs of construction and the continuing costs of operating and maintenance. If they are going to borrow money, they have to be able to pay back the interest. If they are going to charge tolls, they have to compare the amount of the toll to what users would actually save by using their facility. Set the toll too high and nobody will use the facility.

Potential investors have a simpler and more direct question: if they put their money into the project, would they be able to recover their investment plus a reasonable return? They should be worried about the feasibility of the project, the time it could take to complete the project, and the ability of the project to actually generate revenue. They would not necessarily have any interest in the details of the construction or the operation, and they would be comparing this project with completely different options for making money.

Contractors may not care at all about what the project ultimately accomplishes, as long as they are able to complete their portion of the project on time, safely, and on budget. They will be very interested in trying to predict construction costs, choosing the most effective technologies and materials, and in planning and managing the process. They must determine whether the potential profits from the project are worth the risks that they perceive to be associated with the project.

Case Study
Building an Office Tower in Manhattan

"The whole magic of our industry is twofold. One is to build a beautiful building but, more important, it's got to be successful. The only way it becomes successful is if you start collecting rent. The sooner you start collecting rent, the sooner the building becomes more successful. The minute you start collecting rent, all the sins of the father are forgiven. Everything that we've done wrong, they forget – we're all friends again."

Marvin Mass, HVAC Contractor, Worldwide Plaza, **Skyscraper**, p. 306

Consider a real estate developer who is looking for opportunities to create value by constructing buildings. If the estimated value of the newly created space is worth more than the expected development costs, then there is a development opportunity. For an office building, the value will be based upon the leases that can be obtained for the office space. The development costs will include the cost of the land, preparation of the site, design & engineering, construction, and possibly various costs related to the approval process. For example, in return for building a new entrance to a subway station, the developer may be allowed to build more intensively.

In Manhattan and other urban centers, land becomes a very expensive resource, which causes strong economic pressures for intensive development. In very general terms, the value of a building will be proportional to the usable space that it contains, i.e. the space that can be leased to clients. Doubling the size of the building will therefore roughly double the usable space and therefore double the value of the building. On the other hand, development costs are not at all proportional to the size of the building. The price of the land depends upon the local real estate market, not the value of what you intend to build; whether you build a single story warehouse or a 50-story office building, the cost of the land will be the same. Moreover, whether that office building is twenty, fifty or eighty stories tall, it will require access to local streets, a lobby and a roof. Adding stories will, for a large building, simply mean replicating the designs and materials used for one story over and over again. While certain structural components will need to be stronger for a taller building, the added costs will be rather minor for a steel-framed structure.

Since there are economies of scale in building, the incremental cost of adding another story will be well below the average cost, while the incremental value of another floor of leasable space will not diminish (assuming the space can be leased!). Hence, adding more stories and maximizing the usable space on each story will increase the value of the project while reducing the average cost/square foot of the project. The developer therefore has a strong incentive to build the largest possible building.

There are various constraints to the size of the building that can or will be built:

- Zoning regulations may limit the portion of the site that can be developed or the total floor area ratio (FAR, the ratio of floor space to the area of the site).
- Technological capabilities may limit the height (although the limit is obviously more than 100 stories and seldom if ever a real limit today).
- Market considerations may limit the amount of space that the developer wishes to make available today.

Karl Sabbagh, in a highly readable book called "Skyscraper", described the re-development of an entire block in Manhattan during the mid-1980s. The site, which had formerly been occupied by Madison Square Garden, was between 49th and 50th streets and 8th and 9th Avenues, a location in a rather rundown area somewhat west of the prime office locations in Manhattan. Developing the site as an upscale office building was somewhat risky, not because the rents would be lower than in the best locations, but because it might not be possible to rent the space at any price. Bill Zeckendorf, the developer, bought the land for $100 million, but only when he was reasonably sure that the site would be able to attract tenants to what he called the "Worldwide Plaza".

The zoning regulations called for a FAR of 12, which was increased by the city to 14 as a bonus for Zeckendorf's agreeing to make some improvements to the subway station on the site and to provide an acre of open space as part of the project. This provided an opportunity for 1.9 million sq. ft. of usable space, of which 1.5 million was in a 50-story office tower. Zeckendorf expected to be able to lease the space in the office tower at rates of $20-$30/sq.ft./year, with possible increases to $40 in the future. These estimated lease rates were discounted by about $5/sq.ft. from the rents achievable a few blocks toward the other side of Manhattan. At $20/sq.ft., the annual rent would be $32 million for the office tower; at $30/sq.ft. the annual rent would be $48 million.

The estimated costs for the entire project were expected to exceed $500 million, and the costs for the office tower were estimated to be $370 million (Table 1). The basic plan was to use a construction loan to cover the construction costs and to refinance to a 30-year mortgage at a lower interest rate once the building opened. The construction loan would have a high interest rate, because of the risks of delays and overruns in construction and the possibility that it might not be possible to lease all of the space. If all went well, the building space would be leased at favorable rates to long-term tenants, and the lease payments would be more than enough to justify a mortgage sufficient to repay the construction loan.

Table 1 Projected Costs of the 50-Story Office Tower

Cost Element	Estimated Cost
Land acquisition (office tower portion of the site)	$58 million
Preparation of case for development (architects and lawyers)	$5 million
Architects, engineers and borrowing costs	$145 million
Construction cost	$145 million
Project management	$17 million
Total	$370 million

Interest rates and lease rates were the keys to the success of the project. Interest rates were likely to be on the order of 10% or more for the construction loan and on the order of 8% for the mortgage. The costs of the construction loan were included in the estimated cost of the building, but delays and unexpected expenses could lead to higher interest payments. With an interest rate of 10% on the construction loan, the monthly interest on a balance of $370 million would be about $3 million (10% per year/12 months/year)($370 million).

If the construction costs were indeed on the order of $370 million, and if Zeckendorf could obtain an 8% mortgage, then the annual mortgage payments would be approximately $33 million:

Annual payment = $370 million [A/P,8%/,30]

= $370 million (0.0888) = $32.9 million

If the building could indeed be rented at $30/sq.ft., then the $48 million annual revenue would seem to provide enough cash to cover this mortgage payment plus some operating expenses for managing the property. However, if the average lease rate were only $20/sq.ft., then the cash flow would be about the same as the mortgage payment, with no reserve for managing the building. Thus, to have a successful project, it would be essential to complete the project on time and on budget, to secure long-term leases with favorable rates, and to secure long-term financing sufficient to cover the costs of construction.

The building was actually constructed for about $380 million, as there were minor overruns in several areas related to construction or material problems. It was rented at rates of $26 to $32/sq.ft., with the lowest rate going to a major tenant who became a part-owner of the building and committed to leasing 600,000 square feet of space at the outset of the project. The next largest tenant obtained a rate of $29/sq.ft., which was lower than the owners wanted, but it was accepted in the uncertain aftermath of the stock market crash of October 1987, just before the building was ready for occupancy. Smaller tenants paid rates of about $32/sq.ft. By the end of the project, monthly interest costs were close to $3 million and deferring rentals was costing close to $4 million per month.

The building was constructed on a "Fast Track" basis in order to minimize borrowing costs during the construction period and to begin lease payments as soon as possible (Table 2). The projected lease payments were sufficient to justify a permanent mortgage that enabled Zeckendorf to repay the construction loan. Despite the unexpected downturn in the Manhattan real estate market, the building project was successful.

Table 2 Project Timetable for the World Wide Plaza

Event	Date
Site Acqusition	1985
Secure major tenant as co-owner	1985
Ground-breaking ceremony	November 12, 1986
Initial target for making space available to tenants	November 25, 1988
All tenants in the building, working on finishing their space	March 1989
Tenants start to move in	May 15, 1989
Permanent mortgage obtained for the entire project	May 31, 1989

The overall viability of the building depended upon being able to complete construction close to budget without major delays and being able to rent the building at something close to the expected rates. Both of these requirements were met. However, the objectives of the different actors were not based upon the overall perspective:

> "In this particular building you have a pretty characteristic group. You have the architect, who has the design as his main consideration. He wants to put up a monumental building, something everybody is going to see and say, 'Hey, wow! That's great!' It's his entry into posterity. The construction manager, HRH, they're interested in having a building up that they don't get sued over, that's going to stay in place. Each of the individual trades have the same interest as the construction company. The only difference is, each of the trades says, 'I'm only going to do so much. The rest is someone else's responsibility.' So then you have to argue out who's actually doing what part of the interface between the various trades. The consultant is working to represent the interest of the owner. Again, he's after a viable building, something the guy can make money with. He's not investing money to lose it. He also wants to make sure it's sound. I tell you, he has about the same interest as the construction manager."[1]

Other participants would include the banks that provide the construction loan and the permanent mortgage. For them, the project involves providing a large sum of money up front in the hopes of receiving a larger sum a few years in the future (when the construction loan is repaid) or receiving an annuity that will provide a guaranteed return over a much longer period (via mortgage payments over a 30-year period). The banks take care to ensure that the size of the mortgage is limited by the ability of the owners to make mortgage payments; the banks will offer lower interest rates if they are more certain that the project will be successful.

---

[1] Sabbagh, op. cit. p. 199

Case Study
Multiple Internal Rates of Return for a Stadium Project

This case involves a hypothetical example in which there are multiple internal rates of return. The case is structured as the kind of situation that could face a young engineering economist trying to rank alternative projects for senior management. If the analysis is done using net present value, then the potential benefits of the project are clear. But when senior management insists upon using the internal rate of return, all kinds of problems emerge.

You have accepted an internship position that gives you a chance to do what you always wanted to do – work in the front office of a major league team. With your background in project evaluation, you have been asked to calculate the financial potential of the proposed new stadium. At a meeting with the president, the general manager, the VP finance, and several consultants, you learn the general plan. The stadium will be constructed over two years, with a total cost of $250 million. Constructing the stadium will lead to two very large immediate payoffs from selling 20-year leases on corporate boxes and selling real estate around the stadium to various hotel and restaurant chains. Revenue will be about $8 million per year in the first 10 years, after which a major rehabilitation and expansion is planned at cost of $50 million. With a somewhat larger stadium, revenues are expected to rise. Management plans to use all of that revenue plus some additional funds for further development of the area; thus cash flows are expected to be negative for several years before there is another big payout in year 20 when they are able to complete and sell more of the real estate. Table 1 summarizes the expected cash flows.

Table 1   Expected Cash Flows for Stadium Project

Year	Investment	Revenue	Cost	Cash Flow
0	100	0		-100
1	150	10		-140
2		150	2	148
3		150	2	148
4		10	2	8
5		10	2	8
6		10	2	8
7		10	2	8
8		10	3	7
9		10	3	7
10		0	50	-50
11		30	20	10
12		30	25	5
13		30	30	0
14		30	35	-5
15		30	40	-10
16		30	50	-20
17		30	60	-30
18		30	70	-40
19		30	90	-60
20		100	10	90

You create a spreadsheet with all of this information and then set out to determine the internal rate of return. First you try 8%, but that gives a positive NPV of $2.8 million; then you try 12%, but that gives a negative NPV of $4.8 million. You figure the NPV will be zero somewhere in the middle of these two, and you eventually determine that the NPV is zero with a discount rate of 9.85%. You therefore prepare a presentation that concludes that the stadium project has an internal rate of return of "just under 10%". You're quite proud of your work and rush into show your boss your result (Table 2).

Table 2 IRR for Stadium Project is Nearly 10%

Year	Cash Flow	8%	12%	9.85%
0	-$100.0	-$100.0	-$100.0	-$100.0
1	-$140.0	-$129.6	-$125.0	-$127.4
2	$148.0	$126.9	$118.0	$122.6
3	$148.0	$117.5	$105.3	$111.7
4	$8.0	$5.9	$5.1	$5.5
5	$8.0	$5.4	$4.5	$5.0
6	$8.0	$5.0	$4.1	$4.6
7	$8.0	$4.7	$3.6	$4.1
8	$7.0	$3.8	$2.8	$3.3
9	$7.0	$3.5	$2.5	$3.0
10	-$50.0	-$23.2	-$16.1	-$19.5
11	$10.0	$4.3	$2.9	$3.6
12	$5.0	$2.0	$1.3	$1.6
13	$0.0	$0.0	$0.0	$0.0
14	-$5.0	-$1.7	-$1.0	-$1.3
15	-$10.0	-$3.2	-$1.8	-$2.4
16	-$20.0	-$5.8	-$3.3	-$4.4
17	-$30.0	-$8.1	-$4.4	-$6.1
18	-$40.0	-$10.0	-$5.2	-$7.4
19	-$60.0	-$13.9	-$7.0	-$10.1
20	$90.0	$19.3	$9.3	$13.7
Total		$2.8	-$4.3	$0.0

Your boss is rather more subdued that you were, as he understands that 10% is no great rate of return for the crafty men and women who run the major league team. He also notes that the cash flows are weird, with several shifts between positive to negative (Figure 1). He fears that there might be a problem with the IRR that you calculated. He therefore runs the cash flows through his program and quickly obtains what is shown as Table 3. A quick look at the bottom row of this table indicates that the NPV is zero when the discount rate is 2%, suggesting that the IRR is a dismal 2%. On the other hand, the table also supports your calculation, as the NPV is also zero for something a little less than 10%.

Now you and your boss can't go to the CEO and the stadium committee and say that the project is perhaps OK, with an IRR of nearly 10%, except that it may be dismal, with an IRR of only 2%. You need to fix this problem – and you need to fix it fast!

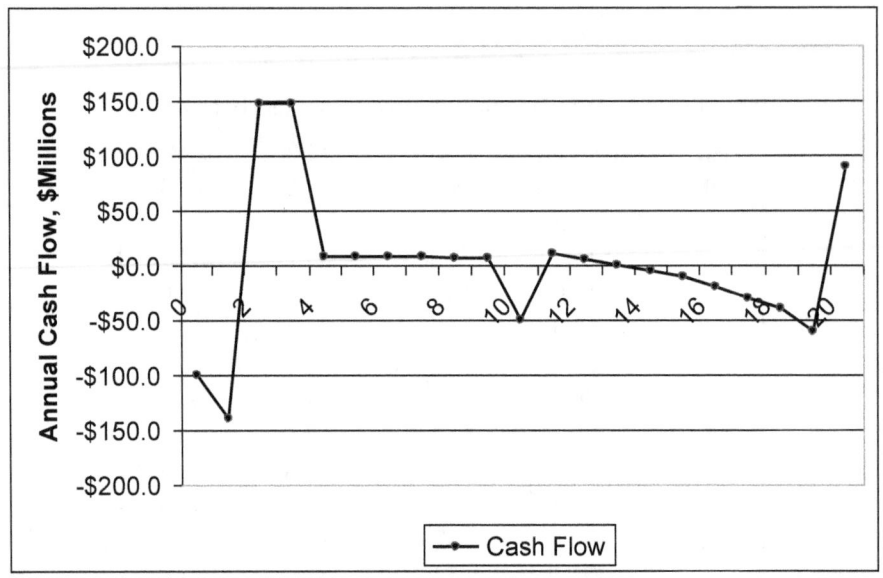

Figure 1 Expected cash flows for the stadium project

Table 3 NPV of stadium cash flows, showing sensitivity of discounted cash flows to discount rate

Year	Cash Flow	Discounted Cash Flows for the Given Discount Rate								
		1%	2%	4%	6%	8%	10%	12%	14%	16%
0	-100.0	-100.0	-100.0	-100.0	-100.0	-100.0	-100.0	-100.0	-100.0	-100.0
1	-140.0	-138.6	-137.3	-134.6	-132.1	-129.6	-127.3	-125.0	-122.8	-120.7
2	148.0	145.1	142.4	136.8	131.7	126.9	122.3	118.0	113.9	110.0
3	148.0	143.6	139.7	131.6	124.3	117.5	111.2	105.3	99.9	94.8
4	8.0	7.7	7.4	6.8	6.3	5.9	5.5	5.1	4.7	4.4
5	8.0	7.6	7.3	6.6	6.0	5.4	5.0	4.5	4.2	3.8
6	8.0	7.5	7.1	6.3	5.6	5.0	4.5	4.1	3.6	3.3
7	8.0	7.5	7.0	6.1	5.3	4.7	4.1	3.6	3.2	2.8
8	7.0	6.5	6.0	5.1	4.4	3.8	3.3	2.8	2.5	2.1
9	7.0	6.4	5.9	4.9	4.1	3.5	3.0	2.5	2.2	1.8
10	-50.0	-45.3	-41.2	-33.8	-27.9	-23.2	-19.3	-16.1	-13.5	-11.3
11	10.0	9.0	8.1	6.5	5.3	4.3	3.5	2.9	2.4	2.0
12	5.0	4.4	4.0	3.1	2.5	2.0	1.6	1.3	1.0	0.8
13	0.0	0.0	0.0	0.0	0.0	0.0	0.0	0.0	0.0	0.0
14	-5.0	-4.3	-3.8	-2.9	-2.2	-1.7	-1.3	-1.0	-0.8	-0.6
15	-10.0	-8.6	-7.5	-5.6	-4.2	-3.2	-2.4	-1.8	-1.4	-1.1
16	-20.0	-17.1	-14.7	-10.7	-7.9	-5.8	-4.4	-3.3	-2.5	-1.9
17	-30.0	-25.3	-21.6	-15.4	-11.1	-8.1	-5.9	-4.4	-3.2	-2.4
18	-40.0	-33.4	-28.3	-19.7	-14.0	-10.0	-7.2	-5.2	-3.8	-2.8
19	-60.0	-49.7	-41.6	-28.5	-19.8	-13.9	-9.8	-7.0	-5.0	-3.6
20	90.0	73.8	61.2	41.1	28.1	19.3	13.4	9.3	6.5	4.6
Total	-$8.0	-$3.3	$0.0	$3.8	$4.4	$2.8	-$0.3	-$4.3	-$8.9	-$13.8

The multiple estimates of IRR occur because the cash flows shift several times between positive and negative. By using the external rate of return method, it is possible to get a better estimate of the return on investment for the stadium project. The costs are all discounted to the present using a rate of 8%, which is the MARR for the stadium owners. The future value of annual revenues is calculated as of year 20 (using the same rate of 8%). The NPV of the costs is $399 million, while the future value FV of the benefits is $1,873 million, as shown in Table 4. The return on investment calculated with the ERR method is the annual rate of return at which $399 million would grow to $1,873 million in 20 years. The answer can be gained by solving the following equation for i%.

$$FV = NPV(1+i\%)^{20}$$

Trial and error on a spreadsheet shows the answer to be 8.04%, so the report to the CEO and the committee could indicate that the project has an expected return on investment of 8%.

Table 4 Calculating the External Rate of Return for the Stadium Project

Year	Investment	Revenue	Cost	Cash Flow	Cost	NPV of Costs	FV of Benefits
0	$100	$0		-$100.0	-$100.0	-$100.0	$0.0
1	150	10		-140.0	-150.0	-138.9	43.2
2		150	2	148.0	-2.0	-1.7	599.4
3		150	2	148.0	-2.0	-1.6	555.0
4		10	2	8.0	-2.0	-1.5	34.3
5		10	2	8.0	-2.0	-1.4	31.7
6		10	2	8.0	-2.0	-1.3	29.4
7		10	2	8.0	-2.0	-1.2	27.2
8		10	3	7.0	-3.0	-1.6	25.2
9		10	3	7.0	-3.0	-1.5	23.3
10		0	50	-50.0	-50.0	-23.2	0.0
11		30	20	10.0	-20.0	-8.6	60.0
12		30	25	5.0	-25.0	-9.9	55.5
13		30	30	0.0	-30.0	-11.0	51.4
14		30	35	-5.0	-35.0	-11.9	47.6
15		30	40	-10.0	-40.0	-12.6	44.1
16		30	50	-20.0	-50.0	-14.6	40.8
17		30	60	-30.0	-60.0	-16.2	37.8
18		30	70	-40.0	-70.0	-17.5	35.0
19		30	90	-60.0	-90.0	-20.9	32.4
20		100	10	90.0	-10.0	-2.1	100.0
Total				-$8.0		-$399.1	$1,873.2

The external rate of return approach is favored by academics, as it avoids the necessity of implying unreasonable returns for reinvesting profits, and it provides a reasonable means of dealing with future periods with negative cash flow. However, this approach is unlikely to be encountered outside of textbooks. Public agencies are apt to consider ratios of benefits to cost rather than ROI, while private companies use the internal rate of return as an easier and apparently more objective result.

Case Study
Public Incentives for Low-Income Housing

Public and private interests may come together to promote multiple objectives by allowing denser development in a suburban setting. Re-zoning land for denser development can offer great opportunities for developers. Requiring the new developments to serve social purposes, such as housing for families with low-income, may be the grounds for a public-private partnership (PPP).

Suppose that a developer is interested in constructing apartment houses in a suburban town where zoning currently allows only single-family housing on 1-acre lots. The developer has plans for constructing three buildings with 10 apartments each on a 5-acre site. The expected cost per unit is $150,000 and the developer plans to lease the units for $2,500/month. Operating expenses are expected to be $500/month per unit. The annual net income is therefore:

Annual net income = 10 units ($2,000 /month/unit) (12 months/year)

= $240,000 per year

Since the ten-unit building is expected to cost $1.5 million to construct, the expected ROI is expected to be $240,000/$1.5 million, or 16%. The developer's MARR is 12%, so this is a very attractive proposition. However, unless the zoning is changed, the 5-acre site will only be able to be used for five single-family houses. Without the zoning change, the developer would have to sell the recently acquired site and seek development rights elsewhere.

The town is interested in creating housing that will be suitable for low- and middle-income families. They thought that it might be possible for the town to build low-income housing, which would be made available to town employees at a maximum rent of $1000 per month. They found that the construction costs for a 5-unit building would be $160,000 per unit, with monthly operating costs of $600 per unit. The net rent per month would therefore be just $400. The town could sell bonds with an interest rate of 4%, so that the annual interest cost per unit would be 4% ($160,000) = $6,400. The net rent of $400 per month or $4,800 per year would be insufficient to cover these interest payments. If the town went this route, they would have to include an additional $6,400 - $4,800 = $1,600 per unit in the town's budget, which they would prefer not to do. The town therefore approached the developer about the possibility of allocating some of the units in the proposed apartments to low-income residents whose rent would be set at $1,000 per month.

The first question is whether or not there is some basis for a partnership. To answer this, we need to determine the maximum reduction in rent that the developer could accept while still earning an acceptable return on the project. With an MARR of 12%, the developer needs an annual return of $180,000 (12% of the $1.5 million investment), which is $60,000 less than the expected rent of $240,000 per year. Reducing the rent from the market rate of $2,500 per month to the desired rate of $1,000 per month would cause a loss of revenue of $1,500/month or $18,000/year for each unit. Thus, even if the developer had to make three units available to town employees at the lower rent, he would still have an acceptable MARR:

ROI with 3 low-income units = ($240,000 – 3 x $18,000)/$1.5 million = 12.4%

If the developed is forced to decide between abandoning the project and accepting a project in which three units are reserved for low-income families paying lower rents, then the developer would likely accept the deal. Of course, the developer would be likely to say "if you provide a subsidy of $1,500 per month per unit ($18,000 per year), you could rent as many as you like."

The town would probably consider $1,600 per year as the maximum subsidy that they would consider, as they could build their own complex if they were willing to provide that level of subsidy. Thus they would be unwilling to provide anything close to the desired subsidy.

On the other hand, they could perhaps offer something else. Suppose the state had approved legislation aimed at promoting the development of low-income housing by allowing the state to guarantee the interest on loans associated with constructing housing in which at least 25% of the units were reserved for qualifying low-income families. Under this legislation, the interest rate on the developer's loans would drop by 2% if the development qualified. If the developer had a loan of $1.5 million, a 2% reduction in the interest rate would be worth more than $30,000 per year. This would be equivalent to $10,000 per unit if three units were reserved for low-income families.

Is this enough to close the deal? Maybe and maybe not. It depends upon how badly the developer wants to proceed and how aggressive the town is willing to be in considering the re-zoning application. It is conceivable that some residents in the town will prefer not to attract low-income families – and it is conceivable that others will be very supportive of initiatives that allow young families and public employees to live in the town. Another step that could be taken would be to allow the developer to add a couple of more units to each building, thereby making the overall development more attractive.

Case Study
Financing a Bridge Project

What is the best way to finance a bridge? Will the state or federal government provide funds, or will the city have to finance the project? Could tolls be sufficient to cover financing costs? If so, will tolls be acceptable to the public? Should the city allow a private contractor to build the bridge and collect the tolls?

Overview of Options for Financing a Bridge Project

A bridge could be built as a public project, a private project, or a public/private partnership (PPP). If the bridge were built as a public project, then there would be several options for financing. The bridge could be viewed as part of the highway system, and whatever funds are used to construct highways could be used to pay for the construction of the bridge. For example, the federal or state government may have a highway trust fund (HTF)[1] that uses income from fuel taxes and registration fees to pay for authorized additions to the highway network (Figure 1). If the bridge is approved as a project that can be supported by the HTF, the design and construction of the bridge can be begun. This is the basic approach that was used in the United States to create the Interstate Highway System and many state highways. State and city governments may also use tax revenues to support highway projects, and they can sell bonds to raise some of the funds required for construction.

Figure 1 Structure of a Highway Trust Fund: money collected from various fees and taxes is used to fund authorized projects, sometimes including transit or intermodal projects as well as highway projects.

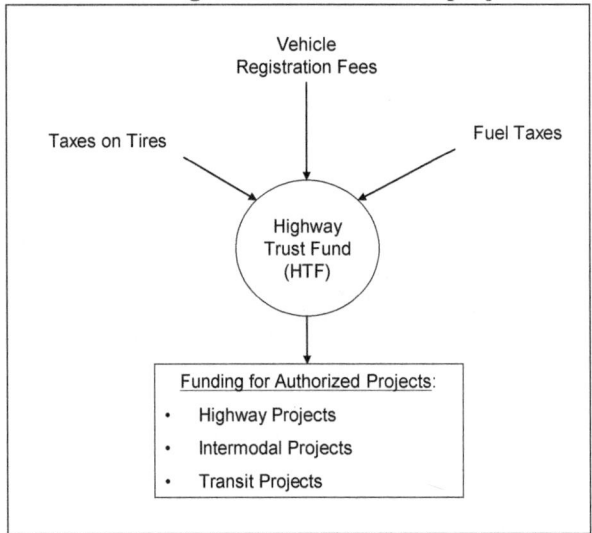

A city or state will commonly finance a bridge project using funds from the HTF or by selling bonds to cover the construction costs and using money from the state's Department of Transportation (DOT) budget to cover annual operating costs. If bonds are sold to help finance the bridge, and if there are no tolls on the bridge, then the bonds are backed by credit of the state or local government. If there are tolls, then the bonds would be backed by the expected

---

[1] The federal HTF was created in 1956 as a mechanism for financing the constructing the Interstate Highway System. Fees and taxes on fuel and heavy trucks provided sufficient revenue to cover the federal government's 90% share of the construction costs. Subsequent legislation allowed small amounts of the fund to be diverted to transit and intermodal projects. The federal fund can only be used for construction, not for operations or maintenance, which remain state responsibilities. States have similar funds, with fuel taxes again providing the major source of revenue. For the complete history of the HTF and the Interstate Highway System, see Tom Lewis, *Divided Highways,* Penguin Books, NY, 1997.

toll payments. Once the bridge is constructed, it clearly belongs to a particular government agency and that agency or another agency is responsible for maintaining and if necessary rehabilitating the bridge. If bonds were sold to pay for the bridge, then those bonds may affect the credit rating of the city or state. If the project were funded out of tax revenues, then it may have been necessary to defer work on schools, water resource projects, or other government projects or activities. Also, some cities or states have limits on the total debt that they can incur. Borrowing to pay for the bridge therefore may limit their ability to borrow for some other purpose. Hence, there may be strong incentives to use tolls to finance the bridge.

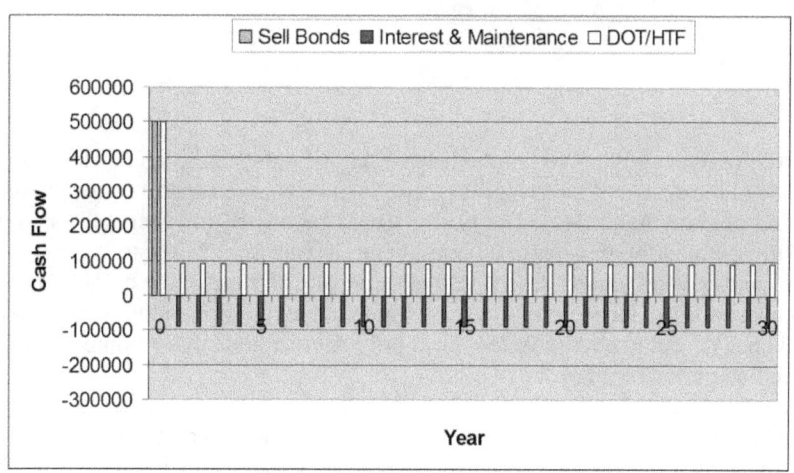

Figure 2 Cash flows from the sale of bonds and money from the state's HTF cover the initial costs of the project, plus continuing maintenance costs and interest payments on the bonds.

Some additional analysis is needed to determine whether toll financing, a state highway project or a PPP is the best option for this bridge. The first question is whether or not the bridge is on a route that would qualify for funding as part of the state highway system. If not, the next question is whether or not high enough tolls can be charged to cover the costs of interest and operations. If so, then the question is whether the city should build the bridge and collect the tolls or whether the city should create a public/private partnership to build and operate the bridge. The city could authorize the bridge and provide the connections to local roads, while a private company could raise funds to pay for the construction costs. The private company would then charge tolls so as to earn a return on its investment. This approach works only if the expected value of the tolls is sufficient to provide an adequate return on capital, e.g. more than enough to cover the interest due on bonds and annual operating expense. The limit on the toll would be the value that users would place upon using the bridge. The toll could be quite high for a bridge that would provide a much shorter or less congested route, but if other bridges are located nearby, then the presence or absence of tolls on those bridges would affect what can be charged on the new bridge. With this approach to building the bridge, the costs would be borne by the private company, not by any public agency, so the construction of the bridge would not affect any public budgets or capital plans. On the other hand, if the bridge is built privately, then the design and capacity of the bridge, as well as the level of tolls charged would be determined by the private company, possibly with an eye toward maximizing profits rather than maximizing public benefits. There could be intense public opposition to allowing a private company to charge what might be viewed as exorbitant tolls in order make excessive profits on an ugly bridge with limited capacity. Thus, there would likely be political pressure to retain some aspect of public control over the project.

Figure 3 Toll Booth on the West Virginia Turnpike: Most of the turnpikes in the U.S. were constructed by state governments prior to the start of the Interstate Highway Program. Tolls cover the interest and redemption costs related to the bonds sold to finance the roads and also cover the costs of maintenance and operations. Tolls can also be used to manage congestion, by charging higher rates at peak hours or by charging higher rates for an express lane. Some states have considered privatizing their toll roads as a means of capturing the value of their investments, either to finance future transport system improvements or to reduce their states' indebtedness.

Various options could be used in a PPP. One common approach is for the public agency to seek bids in which the key variables would include a) the design of the bridge, b) the tolls to be charged and c) the length of time over which the private company would operate the bridge. The bridge would be owned (or eventually be owned) by the public agency, but it would be operated for an extensive period before it was turned over to the public agency. The public role could be to retain control over the size, design, and location, and purpose of the bridge; to ensure that the tolls are reasonable; to provide some financial security for the private company by providing some sort of minimum annual payment if traffic volumes do not rise as expected; or to provide assurance that a competitive project would not be built within some specified period of time.

Let's examine the various financing options using a hypothetical bridge project. Suppose that a new bridge has been proposed that would reduce the travel time and cost between two rapidly growing regions in the rural portion of a state. The bridge, which is strongly supported by local officials, would create a route that would save each user an average of ten miles and 15 minutes. The bridge is expected to cost between $40 and $60 million to construct, and annual maintenance and operating costs are expected to be $4 to $5 million. There are currently 10,000 vehicles per day that use the route, and a preliminary study indicates that nearly all of this traffic would use the new bridge. About 80% of the vehicles on this route are automobiles, while nearly all of the rest are trucks. We will consider three options, namely structuring the project as a routine state highway project, as a private toll bridge, or as a public/private initiative.

Can the Bridge be Justified as a State Project?

Local officials would like naturally prefer the option in which the state pays for the bridge and does not charge a toll, as this would result in the maximum benefits for local citizens and companies. They would also like the bridge to be constructed as soon as possible, preferably within the next two to five years. The big question is whether or not this bridge project can be justified as part of the state's transportation investment plan.

To answer this question, it is necessary to consider the state's transportation budget and the nature of other projects competing for state funds. The state has a prioritized list of transportation projects based upon a formula that recognizes the benefits of reducing congestion, improving safety, reducing travel times, and promoting economic development. For this example, assume that it is apparent that the proposed bridge would not have a very high priority. There are many projects involving bridges and road rehabilitation involving much more heavily traveled routes in more densely populated areas of the state, while the existing route, though long, has very few accidents and essentially no congestion. In short, this is a low priority bridge, and there is no immediate way to dramatize the need for it. Moreover, the state's highway trust fund is substantially underfunded, primarily because fuel taxes have not been increased for nearly 20 years. The trust fund can barely provide enough funds for high priority projects, and medium priority projects have been set back ten or more years in the state's investment plan. In short, local officials cannot expect to have the state pay for the proposed new bridge.

Could a Private Bridge Project be Financed with Tolls?

Would it be possible for a private company to build the bridge? If so, would that be a good idea for the region? The economic value of the bridge is the time and cost saved by those who use the bridge to shorten their travel distances plus additional benefits related to economic development that is likely to result from the increase in mobility provided by the new bridge.

In this example, assume that the major economic benefit comes from a reduction in travel expense for those that use the new bridge. The state DOT estimates the marginal cost per mile for driving an automobile to be $0.20, taking into consideration the cost of fuel and the wear and tear on the vehicle. The marginal cost per mile for driving a truck is on the order of $0.50. The average value of the time saved is on the order of $10 for automobile passengers and $20 for trucks.

Would a private company be able to finance the bridge by selling bonds backed by toll revenues? The first step is to estimate the annual revenue that must be raised by the tolls. If the bridge is financed by selling corporate bonds, the interest rate would have to be about 8%. The interest costs would therefore be 8% of the construction cost, or $3 to $5 million.[2] The total annual cost, including maintenance and operations as well as interest, would therefore be $8 to $10 million.

The next step is to estimate the potential annual revenue. The toll can be no higher than the economic benefit of using the bridge. Using the DOT cost numbers, the average benefits per user can be estimated:

    Auto benefits = 10 miles ($0.20/mile) + 0.25 hours ($10/hour) = $4.50
    Truck benefits = 10 miles ($0.50/mile) + 0.25 hours ($20/hour) = $10
    Weighted average benefits per vehicle = 0.8 ($4.50) + 0.2 ($10) = $5.60

If 10,000 vehicles used the bridge per day, the annual economic benefits would be as follows:

    Annual benefits = $5.60/veh. (10,000 veh./day) (365 days/yr) = $20 million per year.

In other words, the annual economic benefits appear to be at least double the annual costs for interest and operations, even if the bridge costs are at the high end of what is anticipated. A toll of $3 for automobiles and $6 for trucks would be sufficient to cover annual costs. Moreover, since the regions served by the bridge are rapidly growing, traffic volumes and toll revenues would be expected to rise. Thus, it does appear to be feasible for a private company to build the bridge using money raised by selling bonds and paying the interest on the bonds with tolls that users would be willing to pay.

Should the Bridge Be Built as a Public Private Partnership?

The public might well object to the prospect of a private company building the cheapest possible bridge and charging the highest possible tolls. The analysis has shown that a toll of $3 per car and $6 per truck would be more than sufficient to cover the 8% interest rate that the private company would pay on its bonds. However, the same analysis showed that the toll could be nearly 50% higher and still attract most of the traffic. If the bridge is as critical as the local officials believe, and if the region continues to grow as expected, then traffic volume - and toll revenues – would be expected to increase substantially over the life of the bridge. Might it be better to structure the project as a public/private partnership?

---

[2] If the construction cost is at the low end of the estimates, then the annual interest will be 8% of $40 million; at the high end of the estimates, the interest would b 8% of 60 million. These are all estimates, so all that can be said is that the interest payments are likely to be $3 to $5 million per year.

The logic for public involvement is that interest costs could be lowered and that tolls could be controlled. A regional authority could be created that would approve the design for the bridge and own the bridge, and this authority could seek a partner or partners to construct and operate the bridge. With public backing, it would be possible to get lower interest rates by selling tax-free municipal bonds to fund the project. Even if the regional authority were unwilling or unable to sell bonds to finance the project, they could still seek bids for constructing and operating the bridge. They could also stipulate that the bridge (and the toll revenues) would revert to the regional authority after a period of twenty or more years.

Case Study
Using a Probabilistic Model to Investigate Financial Risks

This example shows how to create a simple spreadsheet that can be used to explore risk and uncertainty using random variables for the key factors in the analysis.

When evaluating a potential project, owners and investors will naturally be concerned about making enough profit to achieve their minimum acceptable rate of return. If they are borrowing money for the project, then they want the profit from operations to be large enough to cover interest payments on the debt that they have incurred while also providing an acceptable rate of return on their equity, i.e. what they themselves have invested.

(Eq. 1)   Investment = Equity + Debt

(Eq. 2)   Profit = Revenue – Cost – (Interest rate)(Debt)

(Eq. 3)   ROI = Profit/Equity

In any project, there will, at the outset, be considerable uncertainty related to each of the variables in these equations. The total investment will not be known until construction is completed, and the amount of debt and the interest rate on the debt will not be known until financing is arranged. Actual revenues and operating costs will be uncertain for some time after project completion, as it may take some years for the entire project to reach full operation.

It is possible to treat all of the variables in Equations 1 to 3 as random variables. Each time that a random variable is encountered in an analysis, a value for that variable is generated from a set of possible discrete values or a possible distribution of values. If there are more than one random variables in a set of equations, then they could be independent or dependent. If they are independent, then the value selected for one variable does not affect the value selected for the other variable. If they are independent, then the value selected for one variable will influence the value selected for the other variable.

Random variables can be created in spreadsheets by using the function that generates random numbers. In Excel, for example, it is possible to use the function RAND() to generate a random number greater than or equal to 0 and less than 1.0. Whenever the spreadsheet is recalculated, a different random number will be returned. This function can be used to create random variables that take on a value that is believed to be uniformly distributed between a minimum value A and a maximum value B:

(Eq. 4)   Random Value between A and B = A + RAND() (B-A)

A random variable can also be defined to take on discrete values. For example, suppose that a random variable could take one of two equally likely values V_1 or V_2. This random variable can be created in a spread sheet as follows. First select a random number by using RAND(), then create a scheme such as this:

(Eq. 5)   Discrete Random Variable = V_1 if $0 \leq$ RAND() < 0.5

(Eq. 6)   Discrete Random Variable = V_2 if $0.5 \leq$ RAND() < 1

If the values are not equally likely, then it is possible to adjust the ranges in these equations. For instance, if V_1 is three times as likely as V_2, then:

(Eq. 7)   Discrete Random Variable = V_1 if $0 \leq$ RAND() < 0.75

(Eq. 8)   Discrete Random Variable = V_2 if $0.75 \leq$ RAND() < 1

This process can readily be extended to allow any number of discrete values, each with their own probability, by dividing the interval between 0 and 1 into increments such that the length of each increment equals the desired probability for each value. Thus, we can now create a random variable R that takes of values V_1, V_2, ... V_n with probabilities P_1, P_2, ... P_n.

It is possible to create new random variables as the sum, ratio, or other functions of multiple random independent variables. Thus, if investment, revenue and cost are independent random variables, then profit and ROI will also be random variables. It is also easy to create a random variable that is dependent upon the value of another random variable. This can be done by using statements such as this in a spreadsheet:

(Eq. 9) $R_2 = a + b R_1$

In this case, the value of R_2 will depend upon the value of R_1. More complex equations can also be used to allow different calculations for R_2 depending upon the value of R_1. For example:

(Eq. 10)   If $R_1 < A$ then $R_2 = 100 + R_1$ but if $R \geq A$, then $R_2 = 50 + R_1$

With some ingenuity, it is possible to design quite complex simulations using a spreadsheet. A time-based simulation can be structured by having values of variable in one time period depend upon the situation at the end of the previous time period. If all the variables, random numbers, and calculations are included in one (possibly very long) row of the spreadsheet, then copying that row over a range of rows will produce a simulation of how the system will change over time. The following hypothetical example demonstrates how random variables can be used to estimate the probability that a proposed project will be acceptable.

Suppose a company is considering making an investment of $100 million to purchase the rights to operate an existing toll road. Toll revenues are expected to range from $20 to $25 million annually, while operating and maintenance expenses are expected to range from $10 to $15 million annually. The company plans to sell bonds to raise $80 million at 8% interest and raise $20 million from equity investors who anticipate a 10% ROI. Assuming that revenue and operating costs are independent random variables, what is the likelihood that the ROI will exceed 10% in any given year? What is the likelihood that net revenues in any year will be insufficient to cover the interest costs? A quick glance at the minimum expected revenue and the maximum expected cost shows that the annual revenue could be as low $5 million, which would be insufficient to cover the $6.4 million interest on the bonds. Is this something that investors should worry about?

To answer this question, revenue and cost can be defined as independent random variables. Investment cost, total debt, and interest rate on debt are assumed to be $100 million, $80 million and 8%. Profit will be as follows:

(Eq. 11)   Profit = Revenue – Cost – 8% ($80 million)

Table 1 shows results from a probabilistic model that was created in a spreadsheet. The table has 20 rows, each of which could be viewed as a separate year over a 20 year period or 20 different random results for a single year. Investment, debt and interest rates are assumed to be as given, but the spreadsheet has an area at the upper right where these factors can be modified. Each row of the main body of the spreadsheet has two random variables. The first is used to calculate a value for revenue and the second is used to calculate a variable for cost, both using Equation 4. Profit is calculated using Equation 11 and ROI is calculated as profit/equity (Equation 3). The average ROI is calculated as the average profit divided by the net investment of $20 million. According to this analysis, the financial aspects of the project are solid. There is very little chance that net revenues will be unable to cover costs, and the expected ROI is over 20%, well above what the investors have been promised. The annual values vary considerably, reflecting the uncertainty in the estimates of revenue and cost. However, the risk of failure is very low.

Once this spreadsheet was created, it was possible to conduct further analyses within a few minutes. For example:

- Running the analysis 10 times produced average ROI ranging from 16.8% to 23.2%; the median ROI was 18.1%.
- If interest rates were increased to 9%, the average ROI ranged from 8.7% to 18.4%; the median ROI in 11 trials was 14%. Increasing interest rates on the bonds naturally reduces ROI, but the project still can be expected to have better than a 10% return, which is what the investors are looking for.
- If there is a cost overrun, and the investors have to put up an additional $20 million, the ROI ranged from 6% to 11.4% with a median of 8.6%. Thus, controlling construction costs appears to be quite important.

Table 1 A Spreadsheet That Uses Random Numbers to Explore Risks and Uncertainties Associated with a Project

			Investment	$100	million	
			Debt	$80	million	
			Interest	8%		

Year	Random Number 1	Random Number 2	Revenue	Cost	Interest	Profit	ROI
1	0.887713	0.138114	24.44	10.69	6.40	7.35	36.7%
2	0.997132	0.077517	24.99	10.39	6.40	8.20	41.0%
3	0.651748	0.191994	23.26	10.96	6.40	5.90	29.5%
4	0.474722	0.464287	22.37	12.32	6.40	3.65	18.3%
5	0.922596	0.886441	24.61	14.43	6.40	3.78	18.9%
6	0.415191	0.818556	22.08	14.09	6.40	1.58	7.9%
7	0.739819	0.764117	23.70	13.82	6.40	3.48	17.4%
8	0.425825	0.106793	22.13	10.53	6.40	5.20	26.0%
9	0.771215	0.067184	23.86	10.34	6.40	7.12	35.6%
10	0.590406	0.437435	22.95	12.19	6.40	4.36	21.8%
11	0.607368	0.91192	23.04	14.56	6.40	2.08	10.4%
12	0.005216	0.406998	20.03	12.03	6.40	1.59	8.0%
13	0.382789	0.919088	21.91	14.60	6.40	0.92	4.6%
14	0.003209	0.711268	20.02	13.56	6.40	0.06	0.3%
15	0.479641	0.470464	22.40	12.35	6.40	3.65	18.2%
16	0.172015	0.24961	20.86	11.25	6.40	3.21	16.1%
17	0.648983	0.107807	23.24	10.54	6.40	6.31	31.5%
18	0.065481	0.587854	20.33	12.94	6.40	0.99	4.9%
19	0.930513	0.395585	24.65	11.98	6.40	6.27	31.4%
20	0.94164	0.431054	24.71	12.16	6.40	6.15	30.8%
Total			455.57	245.72	128.00	81.85	20.5%

Case Study
Applying Performance-Based Technology Scanning
to Intercity Passenger Transportation

This case study applies performance-based technology scanning to intercity passenger services, drawing upon research conducted for the UIC (Union Internationale de Chemins de Fer).[1] As the research was conducted for a railway organization, the focus was on how to enhance the competitiveness of rail transportation, and that is the perspective that is taken throughout this section. The research ultimately related to the types of projects that would be desirable for attracting more intercity travelers to the railways of the world. Increasing the maximum train speed is often seen as the best way to improve the competitiveness of passenger rail service, but considering technological options using the PBTS framework suggested many other opportunities for improving rail competitiveness. Construction of high-speed rail corridors will be helpful in some locations, but other types of projects may prove to be just as useful and more cost-effective.

Competition for Intercity Passenger Services

The role of rail in intercity transportation varies widely around the world. In China, India, and Russia, where incomes are low and the rail network is extensive, rail has the largest market share for intercity travel. In much of Latin America, bus is the preferred mode even for distances greater than 600 miles. In the most developed countries, railways must compete with air for longer distance travel and autos for shorter distance travel. Still, when railways are able to offer service on the order of 100 mph along 200-300 mile corridors, they can capture more than half the non-auto market. Examples include Paris-London, Stockholm-Gothenburg, and Rome-Bologna. Where railways offer service in excess of 120 mph, they can dominate such markets, e.g. Paris-Brussels, Paris-Lyons, and Tokyo-Osaka. In the US, passenger rail services are highly competitive only for the Northeast Corridor, and rail market share is very low elsewhere.

Competition for intercity passenger services is based upon cost, time, and quality of the available services, which can be modeled using the economic concept of utility. Travelers' utility can be increased by reducing costs, increasing speed, or improving the quality of their experience. Travelers will choose the mode that allows them to reach their destination with the greatest utility. Technological changes can improve utility for potential customers and therefore increase market share, as demonstrated in this section.

Air, bus and auto are the primary modes competing with rail for intercity passengers. For air, the key factors are the time required at the terminal and the number of stops, as well as the actual flight time. For rail competitive trips (less than 1,000 miles), flight time is at most several hours, often less than half the total trip time. Fares, access time, terminal processing and time and hassle associated with connections are key elements affecting travelers' utility.

Bus is much simpler than air or rail, as the terminal time and amenities are both minimal. The average trip time is dependent upon highway conditions and the number of stops. Travel by bus allows opportunities for work, and seats in the best buses are at least as comfortable as coach class on most planes. Design of bus networks is extremely flexible and readily integrated with air or rail networks.

Auto travel is the most flexible, the most comfortable, and often appears the cheapest. Most people ignore depreciation and treat insurance and taxes as fixed or sunk costs, worrying only about out-of-pocket costs (fuel and tolls for personal automobiles, plus daily fees and mileage fees for rental cars). Auto competition varies greatly across the world, in terms of availability, service, and cost. Where roads are poorly developed or extremely congested, auto is too slow for anything but short trips. Where auto ownership is high and highways well-developed, auto is a convenient, cheap

---

[1] Martland, C.D., Alex Lu, Dalong Shi, Nand Sharma, Vimal Kumar, and Joseph Sussman. *Performance-Based Technology Scanning for Rail Passenger Systems.* UIC/MIT-WP-2002-02. MIT, Cambridge, MA (July 2002).

option for traveling quite long distances. In Europe and Japan, out-of-pocket costs are high because of tolls and fuel taxes. In China, railways are losing mode share to autos and especially buses as the highway network is expanded.

The Utility of Time

Economists use the concept of utility as a means of understanding how people make economic decisions. People are assumed to make choices that maximize their utility, perhaps unconsciously, thus providing a basis for understanding and modeling the way people make choices. In principle, utility can encompass cost, travel time, comfort, and other factors such as safety and security. Surveys and statistical methodologies can be used to develop models of utility based upon user choices or their stated preferences. Assuming that such models exist, we can compare the utility associated with using rail, air or other modes. If utility is identical for two modes, then we would expect travelers to be indifferent to which mode they use. If utility varies with distance, there may be a breakeven distance at which mode shares will be equal. Models that estimate the percentage of travelers that will choose a particular mode are called travel demand models. These models are used extensively in evaluating transportation projects, as the predicted benefits of any transportation project will depend heavily upon the demand for the new facility or service.

Some general insights concerning utility have been gained from research on travel demand:

- Trip times and reliability are important factors in addition to out-of-pocket cost.
- Value of time is related to, but less than, the hourly wage and may depend upon mode or trip purpose.
- Time spent in different activities is valued differently; time spent moving in a vehicle is generally less onerous than time spent waiting in the terminal.
- Ease of access and ease of using the mode are important.
- Time of day, trip purpose, and service frequency affect choice of departure and arrival times.

These results document great variations in value of the time for different groups of people in various activities. A study of intermodal facilities for intercity rail, bus, and transit facilities, suggested using 1/3 of the prevailing wage for the travel time from home to work, 1/6 of the prevailing wage for non-work travel, and 200% of the prevailing wage for work-related travel.[2]

Studies of demand typically use rather general independent variables, often just separating trip time into in-vehicle and out-of-vehicle time. A study of high speed rail conducted for the U.S. Federal Railroad Administration, for example, used trip time, fares, and frequency of service in developing demand models for various market segments.[3] That study addressed the potential markets for high speed rail as an alternative to air and auto travel; it defined service to be total trip time, without attempting to distinguish among the utility associated with different trip segments. However, since passengers make much finer distinctions concerning utility than this, it is necessary to make some assumption about passenger utility as part of PBTS.

The researchers in the UIC study assumed that time and comfort utilities can be expressed in monetary terms and compared directly to fares and other out-of-pocket costs. They then made assumptions concerning utilities for different segments of a trip in order to illustrate the relative importance of these segments and the opportunities for technological improvements. After all, it is clear to anyone who has ever traveled that the time spent in some portions of the trip is very onerous, while the time spent in other portions of the trip may be neutral or pleasant. Saving a few minutes in travel time by introducing faster trains may not be nearly as beneficial as using better information technology to save the same few minutes in terminal processing or providing in-vehicle communications and entertainment to make travel time more productive or more enjoyable.

---

[2] Horowitz, Alan and Nick Thompson, *Evaluation of Intermodal Passenger Transfer Facilities.* Final Report to the U.S. Federal Highway Administration, DOT-T-95-02. U.S. DOT Technology Sharing Program, Washington, D.C. (September 1994).
[3] Federal Railroad Administration,"High-Speed Ground Transportation for America", US Department of Transportation, Washington, DC, September 1997, pp. 5-10

A Preliminary Model of Passenger Utility

To begin, consider how a business traveler with an average billable rate of $100/hour and a salary of $40 per hour might view the various segment of an air trip:

- Drive to airport, including buffer time required because of access unreliability: unproductive time valued at 50% of the average salary or $20/hour
- Process time: standing in lines, checking-in, going through security, and boarding are not only unproductive, but uncomfortable and stressful, so this time is valued at $50/hour
- Extra time at the airport: conceivably useful for shopping, eating, or reading, but likely broken into segments too small to be productive; valued as somewhat better than driving at $10/hour
- Time on the plane resting, eating (peanuts), waiting: similar to the time in the car, probably negative, but at something less than average salary, so this is valued at $20/hour
- Time on the plane having fun: time spent watching a movie, eating (a real meal), or reading a book may be indistinguishable from time spent at home, so some of the time could be considered neutral, i.e. $0/hour
- Time on the plane working: this could be billable time with a positive value of $100/hour

This individual would presumably associate similar utilities with the corresponding segments of a trip by rail, bus or automobile, although the duration of similar segments could be quite different for each mode. If we break the competing travel options into logical trip segments and use consistent values of time for each activity, then we can estimate the utility associated with the various options available for any trip.

Let's begin with a 250-mile trip, a distance long enough for rail to be competitive with auto and short enough to be competitive with air. Tables 1 to 3 give representative inputs for evaluating trip utility. These tables were copied directly from a PBTS model that was created in a spreadsheet. Table 1 shows sample inputs for calculating out-of-pocket costs. To facilitate sensitivity analysis, some of the expense items have fixed and variable components. Air is the most expensive ($289 one-way), automobile is the least expensive ($123)[4], and rail is in the middle ($162). The table also shows the time required to make a reservation, which is not an out-of-pocket expense, but which will affect utility.

Table 2 shows the factors used to estimate total travel time, including access, terminals, and buffers sufficient to cover likely delays. Non-stop air is the fastest, requiring 5.25 hours; rail and auto are nearly an hour longer. Table 3 shows hypothetical values of time that might be reasonable for a business traveler in the United States for the various activities specified in Table 1; the final row shows the value per hour for the extra time gained by using the fastest mode. Most likely, the extra time is a net benefit to travelers at something close to their average value of time. However, it could be more or less. For a business traveler, the extra time might be spent with the client, leading to a higher probability of having a successful meeting. Table 3 therefore shows that the extra time is worth $150/hour, 50% higher than the value of work time for our hypothetical business traveler. Other travelers might have completely different perspectives on the value of this extra time. For a student traveling home for the holidays, extra time on the train might be valuable time to finish an assignment – or it might mean missing the start of a great party. A vacation traveler might lose 2% of the daylight hours available on the beach during the vacation – or gain time to finish up work before relaxing on the beach.

---

[4] Some business travelers are reimbursed for their use of their own automobiles on company business. The travel allowance is likely to be based upon the fully-allocated costs of owning and operating a car, which is on the order of $0.50 per mile. For a 500-mile round trip, a business traveler might therefore be reimbursed $250, which would be $100 more than the variable costs of gas, tolls, and wear-and-tear on the vehicle, estimated in Table 1 as $0.30/mile or $150 for a 500-mile trip. When passenger service was cancelled between Pittsburgh and Harrisburg in 2009, this factor was cited as a major reason for the lower than expected ridership: the rail service was competing with the private service operated by the potential passengers themselves and subsidized by their employers in the form of mileage reimbursement for use of their cars!

Table 1 Calculating Out-of-Pocket Cost for Each Travel Option

	Air Non Stop	Air Via Hub	Train	Auto	Rental Car
Circuity	1	1.2	1.1	1.1	1.15
Distance 1 way	250	300	275	275	287.5
Days at destination	2	2	2	2	2
Reservations (hours)	0.25	0.25	0.25	0	0.1
Cost (1-way)					
Access to station	$4	$4	$4		$4
Fare – fixed	$100	$50	$25		
Fare/mile	$0.50	$0.40	$0.30		
Expenses/trip					$40
Expenses/mile				$0.30	$0.05
Expenses/day					$40
Access to destination	$20	$20	$10	$0	$0
Parking per day	$20	$20	$20	$20	$20
Total Out-of-Pocket Cost	$289	$234	$162	$123	$178

Table 2 Calculating Total Trip Time, by Mode

	Air Non Stop	Air Via Hub	Train	Auto	Rental Car
Time for trip					
Access to station	0.75	0.75	0.5		0.5
Buffer for access unreliability	0.25	0.25	0.2		
Process time	0.1	0.15	0		0.25
Queue time	0.25	0.35			
Available time in station	0.5	1.5	0.25		
Boarding time	0.2	0.4	0.2		0.2
Travel time - fixed	0.75	1.5	0.2		
Travel time - per 100 miles	0.2	0.2	1.25	2	2
Total travel time in vehicle	1.25	2.1	3.64	5.5	5.75
Travel time - work %	75%	75%	75%	0%	0%
Travel time - entertainment %	0%	0%	0%	10%	10%
Travel time - rest & other %	25%	25%	25%	90%	90%
Travel time - work	0.94	1.58	2.73	0	0
Travel time - entertainment	0	0	0	0.55	0.58
Travel time - rest & other	0.31	0.53	0.91	4.95	5.18
Exit time from vehicle	0.2	0.4	0.2	0	0.25
Exit time from station	0.25	0.25	0.1		
Access to destination	1	1	0.5	0.25	0.25
Buffer for access unreliability	0.5	0.5	0.5	0.25	0.25
Total time	5.25	7.65	6.09	6	7.45

Table 3 Hypothetical Value of Time, by Mode and Type of Activity

	Air Non Stop	Air Via Hub	Train	Auto	Rental Car
Reservations	50	50	50	50	50
Time for trip					
Access to station	20	20	20	20	20
Buffer for access unreliability	20	20	20	20	20
Process time	50	50	50	20	50
Queue time	50	50	50	20	50
Available time in station	10	10	10	10	10
Boarding time	50	50	50	50	50
Travel time - work	-100	-100	-100	-100	-100
Travel time - entertainment	0	0	0	0	0
Travel time - rest & other	20	20	20	40	50
Exit time from vehicle	50	50	50	0	0
Exit time from station	50	50	50	50	50
Access to destination	50	50	50	50	50
Buffer for access unreliability	10	10	10	10	10
Extra travel time	150	150	150	150	150

With these detailed inputs concerning travel time and the value of time, it is possible to estimate our traveler's utility for each mode (Table 4). Time is shown as a "disutility" so that it has the same sign as cost – the mode with the lowest disutility is therefore the preferred mode. The quality of time spent traveling is clearly important; ranking the available options in terms of their disutility gives much different results than ranking by either out-of-pocket costs or time. In particular, rail looks much better, because there is extra time for work and less for processing and access. Although rail takes an hour longer, its disutility is less than the disutility of flying. For someone who can work on the train, driving is not a good option. Renting a car, which looks good in terms of direct cost, is by far the worst choice; it takes time to rent the car and it is usually impossible to work in the car, so the disutility of the time is quite high relative to train or plane.

This particular example emphasizes the importance of "work time" to the decision and shows that the cumulative benefits of lower terminal time, easier processing, and greater accessibility help rail relative to air travel (but hurt rail relative to driving your own car). It also suggests a framework for comparing technologies or projects. Any intercity market will have groups of travelers with diverse needs and values. Some people may be able to think effectively when driving, so they may look forward to having several quiet hours in a car. Vacation travelers are concerned with baggage handling facilities – but day trippers are not. Self-employed businessmen undoubtedly view time and costs of travel far more carefully than corporate travelers, whose personal finances are unaffected by their travel choices. The value of terminal services depends upon the expectations of the customer. Hungry students devour fast food, as long as it is cheap and plentiful; wealthy couples en route to a resort prefer to pass an extra hour enjoying a fine meal; a "road warrior" might grab pizza and a beer and check e-mail. The next section considers how passengers in four market segments might respond to various changes in mode or trip characteristics.

Table 4 Hypothetical Disutility of Travel, by Mode

	Air Non Stop	Air Via Hub	Train	Auto	Rental Car
Direct Costs	$289	$234	$162	$123	$178
Reservations	$13	$13	$13	$0	$5
Travel time					
Access to station	$15	$15	$10	$0	$10
Buffer for access unreliability	$5	$5	$4	$0	$0
Process time	$5	$8	$0	$0	$13
Queue time	$13	$18	$0	$0	$0
Available time in station	$5	$15	$3	$0	$0
Boarding time	$10	$20	$10	$0	$10
Travel time – work	-$94	-$158	-$273	$0	$0
Travel time - entertainment	$0	$0	$0	$0	$0
Travel time - rest & other	$6	$11	$18	$198	$259
Exit time from vehicle	$10	$20	$10	$0	$0
Exit time from station	$13	$13	$5	$0	$0
Access to destination	$50	$50	$25	$13	$13
Buffer for access unreliability	$5	$5	$5	$3	$3
Extra travel time	$0	$360	$126	$113	$330
Total travel time disutility	$43	$381	-$58	$326	$636
Total disutility	$344	$627	$117	$448	$820

Estimating Mode Shares

Given the utilities (or disutilities) for each available mode, it is possible to estimate mode shares using what is called a logit model. The mode share for mode j is calculated as follows:

(Eq. 1)   Mode Share = $(e^{-\text{disutility mode j/scale factor}}) / (\Sigma\, e^{-\text{disutility mode k/scale factor}})$

This type of model is commonly used in travel demand studies. If the disutility of two modes is within 5 or 10%, they each have a sizeable market share; if the disutility of one mode is much greater, then it has a very minor share of the market. The scale factor was assumed to be 25% of the average disutility of the mode with the lowest disutility for each market segment. This factor determines how strongly mode shares vary with the relative costs.

The base case for the sensitivity analysis added three market segments to the example from the prior section: general business, vacation, and student. The latter three market segments have values of time that are 50%, 25%, and 10% of the values for the executive considered above. Each market segment was assumed to have an equal number of travelers.

Six cases were investigated in addition to the base case (Table 5). The first two considered airline strategies:

Case 1 – Discount Air Fares: a new carrier enters the market, halving air fares, but doubling processing times. Rail retains more than half the market, because the trip is too short for air speed to make much difference. Since business travelers expect to be productive, the rail option still looks good.

Case 2 – Business Shuttles: major airlines introduce a service aimed at business travelers. Fares match the discount airlines, but processing, queuing and wait times are halved. This service captures more than 90% of the business market. Vacationers also appreciate the time savings; more than half switch to air. Students, still searching for the best deal, divide fairly evenly among the two air modes, rail, and auto. Overall rail market share plummets to 10%.

Table 5 Sensitivity Analysis for Mode Share

	Air Non Stop	Air Via Hub	Train	Auto	Rental Car
Base Case	2%	1%	67%	29%	1%
Discount Air Fares	18%	3%	56%	22%	1%
Business Shuttle	72%	9%	10%	9%	1%
Lower Rail Fares	58%	14%	24%	4%	0%
High Speed Rail	40%	12%	43%	4%	0%
Easy Access	17%	8%	71%	3%	0%
Two Travelers	2%	0%	54%	40%	4%
Easy Rail & Business Shuttle					
125 miles	7%	4%	69%	20%	1%
250 miles	17%	8%	71%	3%	0%
375 miles	35%	12%	51%	1%	0%
500 miles	56%	16%	27%	0%	0%
625 miles	68%	19%	13%	0%	0%

The next three cases address possible rail responses to the business shuttle. Each helps retain market share, with the greatest benefits for this particular example coming from improving access:

Case 3 – Lower Rail Fares: railways respond to the shuttle by cutting fares by 20%. Executives don't even notice the change; the other groups increase their rail mode share to a quarter or a third. Overall, the rail share recovers to 24% of the market.

Case 4 – High Speed Rail: average rail operating speed is 150 mph rather than 80mph. This is more successful than simply lowering fares, and rail is projected to gain 43% of the market. However, a major project would be needed to achieve such high speeds and it is unclear if prices could remain unchanged.

Case 5 – Easy Access: the average speed is again 80mph, but times are halved for rail processing, access, and reservations, while better on-board seating and services increases the value of time by 20% for business travelers. The value of terminal and on-board entertainment time is increased for everyone with more entertainment, retail and culinary opportunities in the stations and better food and services on the train. Executives are assumed to increase their working time from 70 to 80% of the trip time. The results are very strong for the railways, which become dominant in the first three markets and capture a third of the students.

Sometimes a group is traveling:

Case 6 – Two Travelers: travelers share the cost of auto trips or cab rides. The dominant result is to make driving a very good option, with almost all air traffic and more than 20% of the rail traffic diverting to auto. Rental cars also improve, increasing their share from 1 to 4% and becoming a good option for vacationers and students. Clearly, if a family is going on vacation with children, the automobile will look better for even longer distances. Likewise, if three or four people are traveling together on business, then renting a car may look better, particularly if they can conduct some business while driving.

Distance is obviously another key factor for sensitivity analysis, as rail works best for distances that are rather long for highway travel, yet rather short for airlines. "Easy Access vs. the Air Shuttle" was used as the base case. For the 125-mile trip, rail captured 69% and autos took 20% of the market. For the 250-mile trip, the highway modes essentially drop out and direct air flights capture 17% of the market. As distances increase to 625 miles, the rail share drops steadily, while the air share grows. Air travel via a hub is increasingly attractive for the longer distances, as the cost savings become large enough to justify the additional time.

Implications for Carriers and Terminal Operators

The implications of utility analysis are generally well understood. There is value in reducing travel time, in minimizing process time, and in increasing passenger comfort. There is value in providing a variety of ways for travelers to spend their time and their money. Carriers attempt to capture this value by offering premium services at higher prices. First class and business class travelers enjoy quicker check-in, comfortable and productive waiting areas, larger seats and better food – and they are willing to pay a premium of $100-$200 per flight hour for these privileges. This premium is high compared to the coach fare, but not unrealistic when compared to executive salaries or consulting rates. Carriers also advertise their on-board services, including telephones, movies, games, magazines, and shopping opportunities.

Terminal operators may have been slower to understand the importance of time and utility, but they have certainly responded well over the past 10-20 years. New airports feature greatly enlarged shopping opportunities, food courts, fine restaurants, lounges, TVs, internet access, ATMs and other amenities that make waiting time more valuable to the traveler (and more profitable to the terminal owner). Government agencies and airlines are also concerned about airport access, recognizing the importance of time and comfort to the user as well as the costs of the infrastructure. Similar trends have affected some major train stations, which now offer varied retail and dining opportunities

Implications for Project Selection and Project Evaluation

This example shows how markedly different technologies and types of projects can be compared in terms of their potential effects on passengers' utility. The most striking comparisons are among the three generic responses to the business shuttle for the 250-mile trip (Cases 3-5). Lower fares could be interpreted as investments in any of the many technologies that might reduce cost while leaving service and access unchanged. High speed rail is of course a dominant theme in the evolution of rail technology, in rail R&D, and in proposals for rail investment. Easy access relates to entirely different types of projects, including not only improvements in terminal processing, but also improvements in terminal access. For this example, access is somewhat more important than train speed, and much more important than cost reduction. In general, saving time in access and processing or allowing more productive use of time may be more effective – for the customer – than saving time by running faster.

The rail industry and public agencies are well aware of the potential for high-speed rail systems to attract traffic from congested airports and highways, and extensive R&D and investment programs are in place to advance such systems. In the U.S., the "next Generation High-Speed Rail Technology Demonstration Program" was funded at more than $25 million annually in fiscal 2001 and 2002, exceeding the rest of the FRA's budget both passenger and freight R&D[5]. However, as demonstrated in this case study, higher speed is not the only way to reduce travel time or to enhance travelers' utility, and quite different kinds of technologies and projects may be equally effective in enhancing rail competitiveness.

---

[5] Federal Railroad Administration, "Five Year Strategic Plan for Railroad Research, Development, and Demonstration", Chapter 8, U.S. Department of Transportation, Washington D.C., 2002

This PBTS analysis shows that train speed is only one factor, and perhaps a relatively minor factor influencing travelers' decisions. Total door-to-door trip time, the quality of time spent in each portion of the trip, and the opportunity to use the time for enjoyable or profitable activities are all very important factors. Comfortable trains operating over a dense network at reasonably frequent intervals can compete effectively with both air and auto for trips of 100 to 500 miles.

Figure 1 Dublin Train Station

Case Study
Reducing Risks Associated with Grade Crossing Accidents

This case study shows how cost-effectiveness can be used in conjunction with probabilistic risk assessment to determine the best ways to reduce one category of risks of associated with transportation systems.

Hundreds of people are killed each year in grade crossing accidents. A grade crossing (also known as a level crossing in some countries) is where a road crosses a railroad at grade, so that it is possible for a train and a highway vehicle to collide. Accidents may by caused by people who are too sleepy or too drunk to notice that they are approaching a crossing or by people whose car is stuck in traffic while trying to get across the tracks or whose car breaks down on the crossing. A few accidents are caused by malfunctioning of the signals, and a great many are caused by people who ignore the warnings and try to beat the train across the intersection.

The probability of such an accident occurring varies primarily with the density of rail traffic and with the type of protection that is available. At a crossing equipped with flashing lights, the probability of an accident is on the order of 2 per million trains. If there are 20 trains per day, then the expected number of accidents per year would be (20 trains/day x 365 days per year/1 million trains) x (2 accidents per million trains) = 0.0146 accidents per year for such a crossing. For a rail route with 100 such crossings, the expected number of accidents per year would be 0.0146 (100) = 1.46. Although the probability of an accident at any crossing is very low, the likelihood of an accident somewhere along this route is quite high. In fact, such accidents are not uncommon. In the United State, there are more than a quarter million grade crossings and thousands of grade crossing accidents per year.

The second step in estimating risk is to determine the expected consequences if an accident occurs. Even a small passenger train weighs hundreds of tons, so that the consequences of a collision between a train and an automobile are very predictable. The car will be destroyed, any people who fail to get out of the car are likely to be killed or severely injured, and the locomotive may sustain some minor damage. In addition, the engineer and anyone else in the cab of the train will be suffering an emotional shock after a) knowing that the accident was about to happen and b) being completely unable to stop a train in time to avoid hitting someone trying to sneak across before the train arrives. If the train is a passenger train, the major consequence for most passengers will be a delay to the train; passengers might not even notice the impact and will simply wonder why the train stopped.

Accident rates at grade crossings can be reduced by installing flashing lights, putting in crossing gates (arms that automatically come down and block the travel lane when a train approaches), by installing 4-quadrant gates (4 arms block the entire road, so that motorists cannot run around the gate), or by building a bridge. It is even possible to have the entire road blocked, while uniformed personnel ensure that pedestrians do not try to skip across in front of a train, but this expensive solution can only be justified in very unusual circumstances.

Table 1 shows the cost of installing (or upgrading) to each level of protection along with typical accident rates achieved with this type of protection. Note that the accident rate is driven (in this simplified model, but also in reality) by train traffic, not by highway traffic.

Table 1 Grade Crossing Accident Rates

Protection	Cost/crossing	Accident Rate (per million trains)
Signs only	$500	10
Flashing lights	$20,000	2
Gates	$100,000	1
4-quadrant gates	$200,000	0.2
Bridge	$2,000,000	0

Assume that you are the safety officer in a state Department of Transportation, and you have a budget for improving highway safety. You would like to use some of this budget to reduce crossing accidents. You have categorized crossings into the categories shown in Table 1. With this information, you can calculate the cost effectiveness for each strategy in reducing accidents and identify the most cost effective strategies to pursue.

Table 2 Possible Upgrades

Highway Traffic per year	Trains per Year	Base Acc. /year	Current Protection	Possible Upgrade
20 million	100,000	0.1	Gates	Bridge
20 million	100,000	0.1	Gates	4-quadrant
2 million	50,000	0.05	Gates	4-quadrant
200,000	50,000	0.1	Flashing lights	Gates
20 million	5,000	0.01	Flashing lights	Gates
20,000	2,000	0.02	Signs	Flashing lights
20,000	200	0.002	Signs	Flashing lights

The first step is to estimate the effect of the upgrade on accident rates and the number of accidents per year for each category of crossing. The new accident rate per million trains comes directly from Table 1. The expected accidents per year is the product of the new accident rate per million trains multiplied by the number of trains per year. For example, if 4-quadrant gates are installed for crossings with 20 million highway vehicles and 100,000 trains per year, then we can expect the accident rate to drop to 0.2 per million trains, while the expected number of accidents per year will be 0.2 accidents per million trains multiplied by 0.1 million trains/year or 0.02 accidents per year.

The next step is to compute the cost-effectiveness, which is the cost per annual reduction in accidents. This can readily be calculated as the cost of the upgrade divided by the number of accidents avoided. The accidents avoided per year is calculated as the difference between the base and the new number of accidents per year. The cost of the upgrade is shown above in Table 2. The result is the cost per accident avoided. The most cost-effective measures turn out to be:

1) Install flashing lights at crossings with 2,000 trains per year that are currently only protected by signs ($1.25 million per accident avoided per year)
2) Install gates at crossings with 50,000 trains per year that are currently only protected by flashing lights ($2 million per accident avoided per year)
3) Install 4-quadrant gates at the very busy crossings with 100,000 trains per year ($2.5 million per accident avoided per year.

As the safety officer for the state department of transportation, you would still have to determine whether there are more cost-effective strategies to pursue in terms of reducing risks within your state. To do this, you would have to have an estimate of the consequences of grade crossing accidents, so that you could calculate cost-effectiveness in terms of risk reduction rather than in terms of accident reduction. You could then compare strategies for reducing risks at grade crossings to strategies such as adding more policemen to enforce speed limits, requiring seat belts, or upgrading dangerous highway intersections.

Equivalence Factors for Selected Discount Rate

The first section of each table shows the equivalence factors for present value P given future value F assuming i% discount rate over a period of N years (which is denoted as [P/F,i%,N]). The next two sections show the annuity value A given present value P [A/P,i%,N] and future value F given the annuity value A [F/A, i%,N] assuming discount rates of 6% to 12% over periods of 5 to 40 years.

Discount Rate:	6%		
	[P/F,i%,N]	[A/P,i%,N]	[F/A,i%,N]
5	0.7473	0.2374	5.6371
10	0.5584	0.1359	13.1808
15	0.4173	0.1030	23.2760
20	0.3118	0.0872	36.7856
25	0.2330	0.0782	54.8645
30	0.1741	0.0726	79.0582
35	0.1301	0.0690	111.4348
40	0.0972	0.0665	154.7620

Discount Rate:	8%		
	[P/F,i%,N]	[A/P,i%,N]	[F/A,i%,N]
5	0.6806	0.2505	5.8666
10	0.4632	0.1490	14.4866
15	0.3152	0.1168	27.1521
20	0.2145	0.1019	45.7620
25	0.1460	0.0937	73.1059
30	0.0994	0.0888	113.2832
35	0.0676	0.0858	172.3168
40	0.0460	0.0839	259.0565

Discount Rate:	10%		
	[P/F,i%,N]	[A/P,i%,N]	[F/A,i%,N]
5	0.6209	0.2638	6.1051
10	0.3855	0.1627	15.9374
15	0.2394	0.1315	31.7725
20	0.1486	0.1175	57.2750
25	0.0923	0.1102	98.3471
30	0.0573	0.1061	164.4940
35	0.0356	0.1037	271.0244
40	0.0221	0.1023	442.5926

Discount Rate:	12%		
	[P/F,i%,N]	[A/P,i%,N]	[F/A,i%,N]
5	0.5674	0.2774	6.3528
10	0.3220	0.1770	17.5487
15	0.1827	0.1468	37.2797
20	0.1037	0.1339	72.0524
25	0.0588	0.1275	133.3
30	0.0334	0.1241	241.3
35	0.0189	0.1223	431.7
40	0.0107	0.1213	767.1

www.ingramcontent.com/pod-product-compliance
Lightning Source LLC
Chambersburg PA
CBHW080657190526
45169CB00006B/2162